THE PHILOSOPHY OF QUANTUM MECHANICS

D. I. BLOKHINTSEV

THE PHILOSOPHY OF QUANTUM MECHANICS

SPRINGER-SCIENCE+BUSINESS MEDIA, B.V.

ПРИНЦИПАЛЬНЫЕ ВОПРОСЫ КВАНТОВОЙ МЕХАНИКИ

First published in 1965 by the Joint Institute of
Nuclear Research, Laboratory of Theoretical Physics
Translated from the Russian by Express Translation Service

ISBN 978-90-481-8335-7   ISBN 978-94-017-3427-1 (eBook0
DOI 10.1007/978-94-017-3427-1

PREFACE

The present monograph is devoted to the principal problems of quantum mechanics and is based on the conception first stated in my course on 'Fundamentals of Quantum Mechanics'. The scope and purpose of the above course did not allow some principal questions to be brought out as fully as they deserved, and besides, some important points were only very recently developed to a sufficient extent. This refers especially to the analysis of the action of the measuring instrument, whose dual role as an analyser of a quantum ensemble and as a detector of individual events was insufficiently elucidated.

The reader will find that the present monograph is concerned more with theoretical physics than with philosophy. However, I have never separated *Weltanschauung* from science (and particularly theoretical physics) so that the philosophical implications are also discussed, justifying publication in the philosophical series.

In conclusion, I should like to thank the publisher and the translator, whose initiative and effort have made it possible for the book to reach the English-speaking reader.

D. I. BLOKHINTSEV

TABLE OF CONTENTS

CHAPTER I

THE ILLUSION OF DETERMINISM

> "Necessity of this kind does not take us outside the theological view of nature. For science it is largely a matter of indifference whether we follow St. Augustine or Calvin and term it God's will, or the Turks and ascribe it to kismet or fate. In neither case are we tracing the causal chain, and hence in neither case are we any the wiser. Necessity remains an empty phrase, while the situation remains as it was."
>
> ENGELS, *Dialectics of Nature*

It might be said more fully that the illusion of determinism began with the proud though empty boast of Laplace's "Give me the initial data on all particles and I will predict the future of the Universe".

When we recall these resounding words the time of that great 18th-century scientist appears as far off as though it were viewed through the wrong end of binoculars; today we are more humble and very far from the hopes of the mechanistic age. Even the early years of this century, which have destroyed the naive illusions of the 19th century when it seemed that all was clear and that science had run out of problems, seem very remote.

Today we are facing a new frontier, probably a highly revolutionary one, whose outlines appear immense though still largely shrouded in mist.

It may be that this is merely our impression, derived from a desire to forego former principles now clearly outdated, but which we retain solely because we have no new ones and even because many fear the change with its consequent upheaval of our entire way of thought.

The most striking feature of this trend is that it encompasses not only physics but also biology and cosmology; it is capable of having a great influence on our philosophy. Under these conditions it is entirely in order to take a critical view of the bases of the theory.

These bases include the concept of classical determinism. It would be more accurate to say that the doubt is not so much over determinism itself as over unquestioning worship of ideal determinism.

For a long time humanity believed in divine predestination, and afterwards in rigid causal connection. Engels appreciated the philosophical resemblance and narrowness of these viewpoints, while failure to appreciate this affinity has over the Centuries been the cause of tragedy to many outstanding men.

The desire to know the truth is as old as the human race, and this desire must thrust aside any feeling of caution over the possible loss of the security provided by customary concepts.

We must now accept that we cannot ignore the element of games of chance in the behaviour of the Universe; if we were to believe in a God or other guiding hand, we should still have to accept that this God (or the equivalent) has a penchant for games of chance. His Majesty Chance enjoys the explicit indulgence of the Law and makes our events unexpected or improbable; there is even a range of phenomena where he feels himself particularly at liberty, namely atomic and molecular physics.

Boltzmann, a great physicist and materialist, first remarked on this important feature and put forward the ingenious hypothesis that the entire region of the universe accessible to us is the result of a vast fluctuation such as has never been seen before or since.

We must accept that the world is not built in the tidy or simple fashion envisaged by believers in the various metaphysical systems.

To appreciate this concept it is best to begin with the defects in the ideal we hanker for, and this is the starting point of the present book.

The principle of determinism is particularly clear in classical mechanics, which seeks to define completely the future state of a system on the basis of the initial state.

CHAPTER II

CLASSICAL MECHANICS AND CAUSALITY

Classical mechanics is the simplest example of a theory dominated throughout by determinism. Even if we do not learn it at school, we are taught in our first year at university that the laws of classical mechanics allow one to predict the future of a mechanical system whose initial data are available. This appears a trivial truth but in fact contains much that is false or at least debatable. For example, the initial data cannot be specified with unlimited accuracy, so we really have a certain distribution for the initial data, the true values being known only with a certain probability.

This means that we must consider how far this indeterminacy affects our prediction of the state of the system as $t \to \infty$.

Moreover, the system is influenced by unpredictable accidental forces during its motion. Even if these forces are small, their effect may become considerable after prolonged periods. The effects of such random forces cannot be eliminated from the real motion.

Finally, we must ensure that the system will remain isolated during the time for which we are to predict its future. This means that we must specify the boundary conditions at the boundaries of the region within which the motion occurs. All these three aspects are usually ignored in the exposition of classical mechanics.

Before considering the effects of these circumstances on the predictions derived from the laws of classical mechanics, a few words must be said on two methods of describing the motion of systems in classical mechanics. Consider a mechanical system with f degrees of freedom. Its state is characterized by the values $q_1, q_2, \ldots, q_f$ (in short, q) of the coordinates and by the values of the momenta $p_1, p_2, \ldots, p_f$ (in short, p) conjugate with these. The manifold of points q forms the configuration space $R(q)$, while that of the points p forms the momentum space $R(p)$. The name phase space is given to $R(q, p)$, the two spaces taken jointly, $R(q) \times R(p)$. The state of the system in classical mechanics is characterized by a point in phase space.

The first mode of description is that in which we consider a system which at time $t=0$ lies in the zone $q=q_0$, $p=p_0$ of phase space, for which we compute the trajectory $q=Q(t, q_0, p_0)$, $p=\mathscr{P}(t, q_0, p_0)$ in space $R(q, p)$. This trajectory is computed from Hamilton's equations[1,2], which describe the motion in phase space:

$$\dot{q}_s = [H, q_s], \quad \dot{p}_s = [H, p_s] \tag{1}$$

with $s=1, 2, ..., f$. Here $H=H(q, p, t)$ is Hamilton's function, which equals the sum $T+U$ of the kinetic energy $T(p, q)$ and the force function $U(p, q, t)$ characterizing the interactions of the particles mutually and with the external world. In the special case of $\partial U/\partial t=0$ the force function is the potential energy. Also, $T(p, q)$ is quadratic in p. Further $[A, B]$ is the classical Poisson bracket, defined as

$$[A, B] = \sum_{s=1}^{f} \left(\frac{\partial A}{\partial p_s} \frac{\partial B}{\partial q_s} - \frac{\partial B}{\partial p_s} \frac{\partial A}{\partial q_s} \right), \tag{2}$$

so that

$$[H, q_s] = \frac{\partial H}{\partial p_s}, \quad [H, p_s] = -\frac{\partial H}{\partial q_s}. \tag{3}$$

System (1) describes completely the motion of an isolated mechanical system.

The other mode of description is more convenient when the initial data for the system are undefined.

Consider a large number of identical systems in different states. The points representing these are distributed in phase space, and under certain conditions (which we shall not consider in greater detail) we may speak of the density $\rho(q,p,t)$ of these points, so that $\rho(q, p, t) \, \mathrm{d}p \, \mathrm{d}q/(2\pi\hbar)^f$ ($\mathrm{d}p \, \mathrm{d}q=\mathrm{d}p_1 \, \mathrm{d}p_2 ... \mathrm{d}p_f \, \mathrm{d}q_1 \, \mathrm{d}q_2 ... \mathrm{d}q_f$, is an element of volume in phase space, while the factor $(2\pi\hbar)^f$ is taken as unit phase volume) is the number of systems whose coordinates and momenta lie around the point (p, q) at instant t. Since the number of systems is conserved, $\rho(q, p, t)$ must satisfy the equation of continuity in phase space, i.e.,

$$\frac{\partial \rho}{\partial t} + \sum_{s=1}^{f} \left(\frac{\partial \rho}{\partial q_s} \dot{q}_s + \frac{\partial \rho}{\partial p_s} \dot{p}_s \right) = 0. \tag{4}$$

Then (1) and (3) allow us to put this in the form

$$\frac{\partial \rho}{\partial t} + [H, \rho] = 0. \tag{5}$$

On the other hand, the complete derivative of ρ with respect to time is $\dot{\rho} = \partial \rho/\partial t + [H, \rho]$. Therefore Louiville's theorem follows from (5):

$$\dot{\rho} = 0, \tag{6}$$

which shows that ρ is constant along a trajectory.

In this mode of description we are given at $t=0$ not the initial data but the function $\rho_0 = \rho(q, p, 0)$, which essentially describes the distribution of the initial values of p and q; what we seek is the distribution of p and q for $t>0$. This is equivalent to considering the motion of a large set of independent mechanical systems that differ only in their initial data.

A. EFFECTS OF INITIAL DATA

We may avoid obscuring essentials with complicated expressions by considering only the case $f=1$. First we consider the case of free motion, for which $T=p^2/2m$ and $U=0$, m being the mass of a particle. In this case (5) gives

$$\frac{\partial \rho}{\partial t} + \frac{\partial \rho}{\partial q}\frac{p}{m} = 0, \tag{7}$$

whose general solution is

$$\rho(q, p, t) = \rho_0\left(q - \frac{p}{m}t, p\right), \tag{8}$$

in which $\rho_0(q, p)$ is the initial distribution of q and p. A particular case is an initial gaussian distribution:

$$\rho_0(q, p) = \frac{1}{\pi ab}\exp\left(-\frac{(p - p_0)^2}{a^2} - \frac{(q - q_0)^2}{b^2}\right). \tag{9}$$

This has the variances (mean-square deviations):

$$\overline{\Delta p^2} = \overline{(p - \bar{p})^2} = \tfrac{1}{2}a^2 \quad \text{and} \quad \overline{\Delta q^2} = \overline{(q - \bar{q}_0)} = \tfrac{1}{2}b^2.$$

Then, for $t>0$, (8) gives

$$\rho(q, p, t) = \frac{1}{\pi ab}\exp\left(-\frac{(p - p_0)^2}{a^2} - \frac{(q - (p/m)\,t - q_0)^2}{b^2}\right). \tag{10}$$

It therefore follows that the distribution in p for $t>0$ is

$$\rho(p,\,t) = \int_\rho (q,\,p,\,t)\,\mathrm{d}q = \frac{1}{\sqrt{\pi a}}\,\exp\left(-\frac{(p-p_0)^2}{a^2}\right). \tag{11}$$

while that in q is

$$\rho(q,\,t) = \int_\rho (q,\,p,\,t)\,\mathrm{d}p = \frac{1}{\sqrt{\pi}\sqrt{b^2+(a^2t^2/m^2)}}$$
$$\exp\left(-\frac{(q-(p_0t/m)-q_0)^2}{(b^2+(a^2/m^2)\,t^2)}\right). \tag{12}$$

Thus the p distribution is unchanged, while the q one is altered, Δq^2 increasing with time:

$$\overline{\Delta q^2} = \frac{1}{2}\left(b^2+\frac{a^2}{m^2}\,t^2\right) \tag{13}$$

Thus the initial information about the position of a particle is completely lost at a time $t \gg mb/a$; particles differing in spatial position at $t=0$ later become mingled. The information about the position of the particles in space continues to deteriorate as with the passage of time.

According to M. Born[3], this conclusion is applicable to a much broader class of systems, namely quasiperiodic ones, in which we may introduce the so-called cyclic variables φ_s and the corresponding actions I_s (see for example refs. 1 and 2). The Hamiltonian is constant in these variables and is dependent on I alone; Hamilton's equations in these terms state (for one degree of freedom) that

$$\dot{\varphi} = \frac{\partial H}{\partial I} = \omega, \quad \dot{I} = -\frac{\partial H}{\partial \varphi} = 0, \tag{14}$$

so that

$$\varphi = \omega t + \varphi_0, \quad \omega = \text{const.} \tag{15}$$

Here q is expressed in terms of φ and I as a Fourier series:

$$q(t) = \sum_{n=-\infty}^{+\infty} A_n\,e^{in\varphi}, \tag{16}$$

in which the amplitudes A are functions of I. Equations (15) coincide formally with the equations for the free motion of a particle:

$$q = \frac{p}{m} t + q_0, \quad \dot{q} = \frac{p}{m} = \frac{\partial H}{\partial p} = \text{const}.$$

If ω and φ are initially gaussian, the conclusions drawn for free motion apply in their entirety to quasiperiodic motion. Here the role of the momentum (or rather, the velocity p/m) is taken by the angular velocity ω, while the angle φ replaces q. It is clear that the information about the position of the system will be completely lost if the uncertainty in φ exceeds 2π. Eq. (13), in which a represents the uncertainty in ω and b represents the initial uncertainty in φ, shows that this occurs for $t > 2\pi/a$.

We may note that exactly the same situation occurs in quantum mechanics, but with the essential difference that a very narrow initial distribution in q inevitably leads to a particularly rapid broadening of the distribution. Thus, we see that, for aperiodic motion, there is always a time t such that for $t \gg bm/a$ the uncertainty in the coordinate becomes greater than any uncertainty in the initial information as specified via b, and hence destroys the accuracy of that information.

The information on the position is completely lost for $t \gg 2\pi/a$ in the case of periodic motion.

B. RANDOM FORCES

In the previous section we discussed the effects of uncertainty in the initial data on the future predictability of the state of a mechanical system. However, during its motion a particle may additionally be subject to forces that cannot be predicted exactly. Such forces are inevitable under real conditions, although they may be extremely small. For example, they may arise from collisions between the body and the molecules of the medium, from turbulence or from any other random inhomogeneity in the medium.

It is impossible to give a theory of these effects for the general case, although they have been the subject of searching investigation by mathematicians. However, it is not too difficult to allow for the effects of these forces if they are small relative to the principal forces or if the main motion is described by linear equations. This is the type of case we shall consider below.

We put the equation of motion in the Newtonian form

$$m\ddot{q} = F(q, t). \tag{17}$$

and assume that the force $F(q, t)$ may be resolved into a regular part $F(q)$ and a random part $f(t)$, the latter being dependent on time alone.

We also resolve the coordinate q as $Q + \Delta q$, in which Δq is the deviation caused by the random force f. Neglect the higher powers of Δq in (17) and we obtain a linear equation for Δq:

$$\Delta\ddot{q} - \frac{1}{m}\frac{\partial F}{\partial q}\,\Delta q = \frac{1}{m}f(t). \tag{18}$$

This equation may be solved via a Green's function $G(t-t')$, which fits the equation

$$\ddot{G}(t - t') - \frac{1}{m}\frac{\partial F}{\partial q}\,G(t - t') = -\delta(t - t') \tag{19}$$

and is subject to the boundary condition

$$G(t - t') = 0 \quad \text{for} \quad t - t' < 0.$$

We use Green's function to seek a solution of (18) in the form

$$\Delta q = \frac{1}{m}\int_0^t G(t - t')\,f(t')\,\mathrm{d}t'. \tag{20}$$

The following notation is used for the mean of the product of the random forces acting at different instants t and t':

$$a^2 \cdot \Delta(t - t') = \overline{f(t)\,f(t')}, \tag{21}$$

in which a^2 is a dimensional coefficient. This quantity is essentially the value of the autocorrelation function of $f(t)$ for the two instants t and t'. From (20) and (21) we have $\overline{\Delta q^2}$ as

$$\overline{\Delta q^2}(t) = \frac{a^2}{m^2}\int_0^t \mathrm{d}t'\int_0^t \mathrm{d}t'' \cdot G(t - t') \cdot G(t - t'') \cdot \Delta(t' - t''). \tag{22}$$

We assume that the autocorrelation function $\Delta(t)$ decreases rapidly, becoming essentially zero for $|t| > \tau$. If the Green's function varies smoothly in the interval up to τ, we may integrate with respect to t'' and normalize $\Delta(t)$ via

$$\int_{-\infty}^{+\infty} \Delta(t' - t'')\, dt'' = 1,$$

the result then being

$$\overline{\Delta q^2(t)} = \frac{a^2}{m^2} \int_0^t dt'\; G^2(t - t'). \qquad (23)$$

We apply this formula to the linear equation

$$\ddot{q} + \lambda \dot{q} + \omega^2 q = \frac{1}{m} f(t), \qquad (24)$$

in which λ is a frictional coefficient, ω is the natural frequency, and $f(t)$ is the random force. The Green's function for this equation is [4]

$$G(t) = \frac{\exp\left(-(\lambda/2)\, t\right)}{\sqrt{\lambda^2 - 4\omega^2}} \left(\exp\left(t \sqrt{\frac{\lambda^2}{4} - \omega^2} \right) \right.$$

$$\left. - \exp\left(-t \sqrt{\frac{\lambda^2}{4} - \omega^2} \right) \right) \quad \text{for} \quad t > 0 \qquad (25)$$

and

$$G(t) = 0 \quad \text{for} \quad t < 0. \qquad (25')$$

Here we may consider two cases.

a) *Free motion with friction:* $\omega = 0$, $\lambda \neq 0$. From (23) and (25), for $t \to \infty$

$$\overline{\Delta q^2(t)} = \frac{1}{\lambda^2} \frac{a^2}{m^2} \int_0^t [1 - e^{-\lambda(t-t')}]^2 \, dt' = \frac{a^2}{\lambda^2 m^2}\, t + \cdots. \qquad (26)$$

b) *Harmonic oscillator with friction:* from (23) and (25), for $t \to \infty$

$$\overline{\Delta q^2(t)} = \frac{a^2}{2\lambda m^2 \omega^2} + \cdots. \qquad (27)$$

The corresponding results for zero friction are

$$\overline{\Delta q^2(t)} = \frac{a^2}{3m^2}\, t^3 + \cdots \tag{26'}$$

$$\overline{\Delta q^2(t)} = \frac{a^2}{2m^2\omega^2}\, t + \cdots . \tag{27'}$$

These examples show that random forces may have pronounced effects on the predictability of the motion in classical mechanics.

C. BOUNDARY CONDITIONS

It is not usual to speak of boundary conditions in mechanics; rather, the conditions of isolation of the system are considered. By this is meant that the system is subject to no forces other than those appearing in the equation of motion, or that these do not act on the system during the time interval of interest to us.

In effect, it is assumed that a surface forms the boundary of the region of space in which the motion occurs, this surface being impenetrable to external fields or bodies. Figure 1 shows such a surface for the case

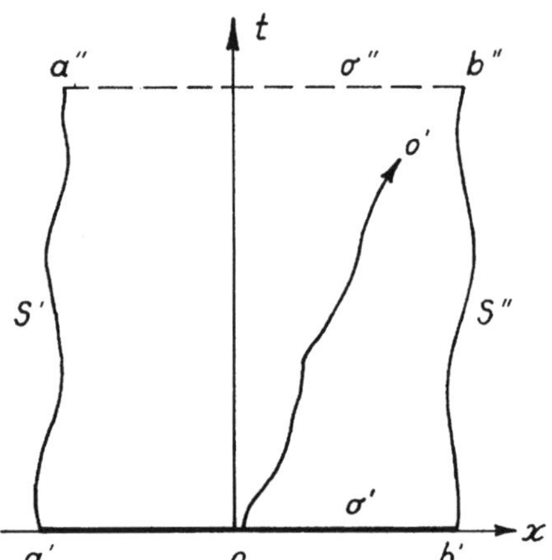

Fig. 1. Initial data are given for the surface σ' (part $a'b'$), while the boundary data are given for surfaces S' and S'' (parts $a'a''$ and $b'b''$). The line OO' is the trajectory of the particle.

of one-dimensional motion along the OX axis. The system is isolated if we can ensure that no external field or body enters through the surfaces S' and S''. For example, a planetary system may be treated as closed only because we can be sure that no unforeseen heavenly body will enter it or pass near it within the period of interest to us.

This feature is extremely important, because it shows that a deterministic prediction of the future is itself conditional: the future of a mechanical system may be predicted only if we can be sure that the system is isolated. The guarantee required here is not implied by the equations of motion but is an additional condition, which produces a great reduction in the reliance on determinism. A vast and depressing "if" arises in the way of the prophet who sets out to predict the future of a real mechanical system.

D. SOME REMARKS ON FIELDS

Consider now the same problem as regards a field that obeys a linear equation; for simplicity, let us limit ourselves to a scalar field $\varphi = \varphi(\mathbf{x}, t)$, which obeys

$$\frac{\partial^2 \varphi}{\partial t^2} - \nabla^2 \varphi - \kappa^2 \varphi = 0. \tag{28}$$

The general solution to this may be expressed in terms of the initial data $\varphi(\mathbf{x}, 0)$ and $\partial \varphi(\mathbf{x}, 0)/\partial t$, together with the boundary values of $\varphi(\mathbf{x}, t)$ and $\partial \varphi(\mathbf{x}, t)/\partial n$, in which n is the normal to the bounding surface, these being given for all future times $t > 0$ by Kirchoff's formula[5]:

$$\varphi(\mathbf{x}, t) = \int \left[g(\mathbf{x}, t; \mathbf{x}', t') \frac{\partial \varphi(\mathbf{x}', t')}{\partial n'} - \frac{\partial g(\mathbf{x}', t', \mathbf{x}, t)}{\partial n'} \varphi(\mathbf{x}', t') \right]$$

$$d\sigma' = \int G(\mathbf{x}, t, \mathbf{x}', t') \cdot \varphi(\mathbf{x}', t') \, d\sigma', \tag{29}$$

in which $g(\mathbf{x}, t, \mathbf{x}', t')$ is the Green's function of (28) that satisfies the inhomogeneous equation

$$\frac{\partial^2 g}{\partial t^2} - \nabla^2 g - x^2 g = -\delta(t - t') \, \delta(\mathbf{x} - \mathbf{x}'), \tag{30}$$

11

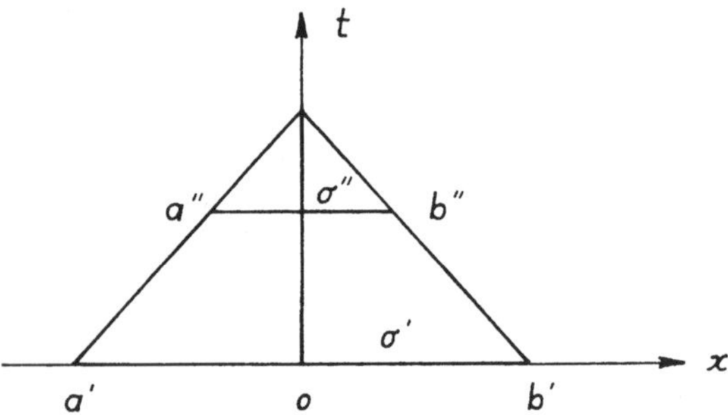

Fig. 2. Initial data provided only for σ' $(a'b')$, with boundary data unknown; the region σ'', within which the field φ is known, contracts with the velocity of light c, and $\sigma'' = 0$ at $t = a'b'/c$.

and G is the operator $g(\partial/\partial n') - \partial g/\partial n'$. The surface σ' over which the integration is carried out has n′ as its normal and consists in part of the time sections S' and S'' (Figure 1) and in part also of the spatial section σ' for which the initial data are given. Formula (29) predicts the values within the volume bounded by the surfaces σ', σ'', S', and S''.

Formula (29) shows directly that the field is expressed not only in terms of the initial data (on surface σ') but also via boundary values (on surfaces S' and S''), which we are obliged to specify for all future time. Then the "if" that arises in mechanics acquires in field theory a direct mathematical expression in terms of the integrals over the time surfaces S' and S''. The physical significance of these integrals is obvious: they describe the effects of fields arising from regions of space not included in the initial data. If we refuse to consider the initial boundary data, our reward is to be restricted (as regards prediction of the field) to a region of space steadily contracting with the passage of time, as shown in Figure 2. Unforeseen signals penetrate into our region of space with the velocity of light, and that region contracts at the same rate; in a time $t = R/c$ the entire region is filled with waves not included in our calculation, and by this time our prediction will have lost all validity.

The linear character of the field allows us to write down directly the cross-correlation of the fields at points $\varphi(\mathbf{x}', t')$ and $\varphi(\mathbf{x}, t)$ if the initial data and boundary values fluctuate. We have from (29) that

$$\overline{\varphi(\mathbf{x}', t')\, \varphi(\mathbf{x}, t)} = \int G(\mathscr{P}, \mathscr{P}'')\, G(\mathscr{P}', \mathscr{P}''')\, D(\mathscr{P}'' - \mathscr{P}''')\, d\sigma''\, d\sigma''', \qquad (31)$$

in which

$$D(\mathscr{P}'' - \mathscr{P}''') = \overline{\delta\varphi(\mathscr{P}'')\,\delta\varphi(\mathscr{P}''')}, \tag{32}$$

is the correlation function for the fluctuations $\delta\varphi(\mathscr{P}'')$ and $\delta\varphi(\mathscr{P}''')$ at points $\mathscr{P}''(\mathbf{x}'', t'')$ and $\mathscr{P}'''(\mathbf{x}''', t''')$ lying on the boundary of the region of integration (on surfaces S', S'', σ'). This expression will not be considered for detailed cases, as it is closely similar to result (22) for mechanics.

Consider the particular case in which the fluctuations are confined to the initial data. There is then a more direct way of calculating the field variance. We expand $\varphi(\mathbf{x}, t)$ as a series in terms of the proper modes $\psi_k(\mathbf{x})$:

$$\varphi(\mathbf{x}, t) = \sum_k q(t)\,\psi_k(\mathbf{x}). \tag{33}$$

The amplitudes of the partial modes $q_k(t)$ are determined by the oscillator equation

$$\ddot{q}_k + \omega_k^2 q_k = 0, \tag{34}$$

in which ω_k is the frequency of mode k:

$$q_k = a_k \exp(i\omega_k t).$$

The cross-correlation function for points $\mathscr{P}(\mathbf{x}, t)$ and $\mathscr{P}(\mathbf{x}', t')$ is

$$
\begin{aligned}
D(\mathscr{P} - \mathscr{P}') &= \overline{\varphi(\mathbf{x}, t)\cdot\varphi(\mathbf{x}', t')} \\
&= \sum_k \sum_{k'} \psi_k^*(\mathbf{x}')\cdot\psi_k(\mathbf{x})\cdot\overline{a_k^*,\,a_k}\,e^{-i\omega'_k t' - i\omega_k t}.
\end{aligned} \tag{35}
$$

If the individual modes are statistically independent,

$$\overline{a_k^*,\,a_k} = \alpha(k)\,\delta(k - k') \tag{36}$$

and the expression becomes

$$D(\mathscr{P} - \mathscr{P}') = \sum_k \psi_k^*(\mathbf{x})\,\psi_k(\mathbf{x})\,\alpha(k)\,e^{i\omega_k(t - t')}, \tag{37}$$

and if $d(k)$ is constant (if the mean of $a_k^* a_k$ is the same for all k), we have instead of (37) that

$$D(\mathscr{P} - \mathscr{P}') = \alpha\sum_k \psi_k^*(\mathbf{x}')\,\psi_k(\mathbf{x})\,e^{i\omega_k(t - t')}.$$

This expression coincides with the Green's function for (28), and for $t = t'$ we obtain that

$$D(\mathscr{P} - \mathscr{P}') = \alpha\cdot\delta(\mathbf{x}' - \mathbf{x}). \tag{38}$$

13

The variance (dispersion) given by (37) does not increase with time. This is a feature distinctive of linear systems.

There are some major difficulties in the discussion of nonlinear systems, but we can obtain some idea of the behaviour of a nonlinear field if we bear in mind that the latter may be considered as an infinitely large set of coupled oscillators. As a system containing a large but finite number of coupled oscillators is quasiperiodic, the variance will increase with time if the initial data are not completely defined. If this conclusion may be transferred to an infinite set, we would expect the variance to increase for a nonlinear field.

E. CONCLUSIONS

We have seen that the basic assumption of classical mechanics (that it is possible to determine uniquely the future state of a system from its initial data) is based on an abstraction that excludes all randomness. The effects of randomness cannot be overlooked in the general case, since the uncertainty arising from the initial data increases with time, and the prediction becomes quite meaningless after the lapse of a certain finite time.

In fact, it is common knowledge that the input data must from time to time be corrected even in a science as precise as celestial mechanics, in order to eliminate cumulative errors.

No machine can be left to operate for an indefinite time without touching the controls to remove errors accumulating during the operation.

The supporters of inflexible determinism are wrong in taking classical mechanics as their example: the motion predicted by classical mechanics is unstable either on account of small random deviations in the initial data or owing to the presence of random forces. No matter how small these perturbing effects may be, there always comes a time at which their consequences will predominate. This instability against slight randomness completely destroys the illusion of unambiguous prediction of the future from the initial data without subsequent correction during the process.

REFERENCES

1. E. T. Whittaker, *Analytical Dynamics*.
2. A. Sommerfeld, *Atomic Structure and Spectral Lines*, transl. H. L. Brose, 3rd ed., 1934.
3. M. Born, *Usp. fiz. nauk* **69** (1959) 173.
4. E. Madelung, *Mathematical Techniques in Physics*.

14

CHAPTER III

A GIBBS ENSEMBLE

Experiment shows that a system subject to random factors may show well-defined laws when the observations are repeated many times. Such laws are usually termed statistical. Random factors are always present in any real case, but their magnitude depends very much on the nature of the system. A system with numerous interacting degrees of freedom may be dominated in behaviour by the random factors; then a dynamic law becomes the slave of chance (a deterministic law is often termed "dynamic" in mechanics), but chance itself produces a new type of law, the statistical regularity.

An example of this occurs in the Maxwell-Boltzmann kinetic theory of matter. Major gifts from His Majesty Chance are the Maxwell distribution and temperature, which relate molecular processes to macroscopic ones. In his H-theorem Boltzmann attempted to show that Chance, strictly speaking, cannot do other than give rise to the Maxwell distribution.

There are no methods for "deducing" statistical laws from deterministic ones; at best they can be made compatible. Any system where chance plays a major part always requires the use of special assumptions of probability – theory type in order to derive its laws, e.g. the assumption of elementary disorder or some other assumption of equal probability for states of a dynamic system.

We shall not enter deeply into these topics, by which learned men have long been confused, and shall assume *de facto* that chance can give rise to a law no less than can determinism.

Gibbs, the founder of molecular statistics, was the first to perceive that we are not obliged to seek the ways in which chance brings a mechanical system to a state that is definite in the statistical sense; we may make some general assumptions and later compare them with experiment.

Gibbs introduced the idea of an ensemble of systems, which has proved of extreme importance throughout statistical physics.[1] Consider a large number of identical and completely mutually isolated systems, each of which interacts weakly with a large macroscopic system at a set temperature (thermostat). This set of systems is known as an ensemble; Figure 3 shows such an ensemble in the form of a row of thermostats $\mathcal{M}$ receding to infinity, together with the systems μ interacting with them. In fact, it is not essential to have numerous thermostats $\mathcal{M}$; one will suffice, provided that the systems that interact with it thereby exert no mutual influence nor alter the macroscopic state of the thermostat.

We now measure some of the dynamic variables (e.g. the momenta p and coordinates q) on a large number of the systems constituting this

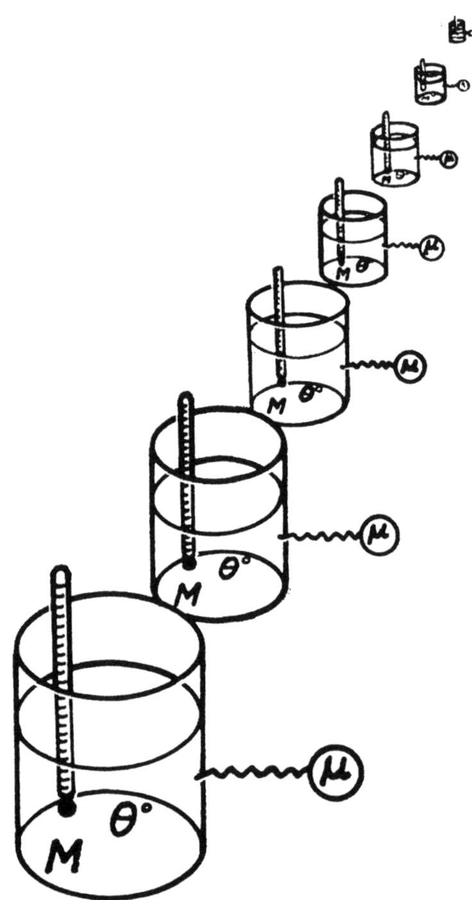

Fig. 3. A Gibbs ensemble in the form of an infinite sequence of identical microsystems μ interacting weakly with thermostats all at the same temperature $\theta°$.

16

ensemble. Gibbs asserted that there is a completely definite probability for any given result from this experiment, i.e. that we have the canonical distribution

$$W_\theta(p, q)\, \mathrm{d}p\, \mathrm{d}q = \exp\left(\frac{\psi(\theta) - \varepsilon(p, q)}{\theta}\right) \mathrm{d}p\, \mathrm{d}q, \tag{1}$$

in which $W_\theta(p, q)\mathrm{d}p\, \mathrm{d}q$ is the probability of finding that the momentum lies in the range p to $p + \mathrm{d}p$ and that the coordinate lies in the range q to $q + \mathrm{d}q$ (all formulae are explicitly for one degree of freedom, for simplicity); θ is the temperature of the thermostat, $\varepsilon(p, q)$ is the energy of the system, and $\exp[\psi(\theta)/\theta]$ represents a normalization factor.

The existence of this distribution is a hypothesis; on this point Gibbs[1] stated that the distribution represented by (1) is the simplest conceivable case, because it has the feature that the law of distribution between phases for the parts is the same for all when the system consists of parts with distinct energies, this feature very greatly simplifying the study and forming the basis for extremely important relationships in thermodynamics. This view may be supported by the following argument[2]. Louiville's theorem indicates that the density $\rho(p, q)$ of points in phase space is constant along a trajectory, which means that it must be a function of the integrals of motion alone. We now assume that our system can be divided into two weakly interacting parts A and B; then ρ is the product of the densities for subsystems A and B:

$$\rho_{A+B} = \rho_A \cdot \rho_B. \tag{2}$$

This means that ρ can be a function of the additive integrals of motion alone. If we disregard macroscopic movements of the system as a whole (rotation, translational motion), the only additive integral of motion is the energy of the system:

$$E_{A+B} = E_A + E_B.$$

Solution of the functional relation of (2) now gives

$$\rho = e^{\alpha + \beta E}. \tag{3}$$

Bearing in mind that only few systems will have infinitely large energy, we conclude that $\beta < 0$; let $\beta = -1/\theta$. Also, α is defined by the normalization condition

$$\int \rho(p, q)\, \frac{\mathrm{d}p\, \mathrm{d}q}{2\pi\hbar} = 1. \tag{4}$$

Hence $\alpha = \psi(\theta)/\theta$. Next we assume that one of the systems is very large, so that it may be considered as a Gibbs thermostat; comparison with (1) then shows that θ is the temperature characterizing the ensemble in equilibrium.

An attempt is sometimes made to derive the Gibbs distribution by considering microcanonical ensembles, which involves fresh hypotheses. To avoid the tangled question of axiomatics, we shall consider only the original approach derived from Gibbs, because this is sufficient for our purpose. An exceptionally interesting feature of $W_\theta(p, q)$ is that it relates the parameters of the macroscopic environment $\mathcal{M}$ (i.e. the temperature θ of the thermostat) to those of the microsystems μ (momenta p and coordinates q). We may say that $W_\theta(p, q)$ relates to a microsystem in a specified macroscopic environment. A Gibbs ensemble is merely the result of bringing together a large number of such situations; in many cases it is practically feasible to a high degree of approximation.

For the sake of future requirements it should be pointed out at this point that no great significance is attached to the question whether $W_\theta(p, q)$ relates to a single particle or is a characteristic of many particles (a question posed largely by the disputes of believers). The essential point is that, no matter whether the observation is made, a single measurement alters only our subjective relation to the observed fact: if a typical phenomenon occurs it is reasonable to say "as expected", while if a rare event occurs we may merely express our surprise or pleasure, as at a good win in a lottery. All of this concerns subjective evaluations; the only thing of objective significance is the distribution of the results of measurements, and this is derived from the performance of numerous measurements on the ensemble, this being the distribution predicted by the probability $W_\theta(p, q)$.

The existence of this probability is one of the striking and fundamental laws of statistical molecular physics.

Here once more we are indebted to the game of chance, which conflicts with its one intrinsic nature in giving rise to laws as simple as the canonical Gibbs distribution.

REFERENCES

1. J. W. Gibbs, *Principles of Statistical Mechanics.*
2. L. Landau and E. Lifshits, *Statistical Physics* [in Russian], ch. 1, GITTL, 1951.

CHAPTER IV

A QUANTUM ENSEMBLE

Consider the case of a Gibbs ensemble whose thermostat is at zero temperature. The canonical distribution implies that the energy of the system must be as low as possible, which means that the kinetic energy $T(p)$ must be zero, while the potential energy $U(q)$ must be at the minimum. This further implies that the momenta must be zero, and the coordinates are obliged to have the one single value corresponding to the minimum potential energy. In other words, $\theta = 0$ implies that each dynamic variable is confined to a single value, with no statistical spread at all. In particular, there should be no light scattering due to molecular thermal motion. However, experiment shows that such scattering does occur at low temperatures and in the limit (at absolute zero). This shows that motion persists in some form at this temperature, where all motion should have ceased. James, Brindley, and Wood were among the first to examine this, and in their paper[1] on the scattering of X-rays in an aluminium crystal they stated that it is necessary to assume a zero-point energy (i.e. an energy of motion at absolute zero) in order to obtain agreement with experiment.

This unfreezable motion is a motion of a new (quantum) type. A Gibbs ensemble spontaneously becomes a quantum ensemble as absolute zero is approched. The problem here is then whether any new laws arise in consequence, and if so, what is their form. It turns out that there exists the full symphonic of the new statistical laws governing the motion of microparticles when macroscopic bodies, which determine the conditions of motion for microparticles, are at absolute zero. These new laws constitute quantum mechanics, which deals with the motion of microparticles in a quantum ensemble.

A very important characteristic feature of a quantum ensemble is that there is a fundamental relation between the variance $\overline{\Delta x^2} = \overline{(x - \bar{x})^2}$ of the coordinate x (in which the bar denotes an average over the ensemble, $\bar{x}$ being the mean of x for the particle or microsystem) and the variance $\overline{\Delta p^2} = \overline{(p - \bar{p})^2}$ of the conjugate momentum p. This is known as Heisenberg's uncertainty principle

$$\overline{\Delta p^2} \cdot \overline{\Delta x^2} \geqslant \frac{\hbar^2}{4}, \tag{1}$$

in which $\hbar$ is Planck's constant.[2,3]

So far the arguments have been general. Let us now turn to a more rigorous definition of a quantum ensemble. Consider a set of macroscopic bodies, which can be termed the macrosetting $\mathcal{M}$. This macro environment in some way determines the state of motion of the microsystem μ (sometimes simply the word "state" will be used, the word "motion" being omitted). We now imagine $\mathcal{M}$, with its μ, reproduced an indefinitely large number of times, as for the thermostates and microsystems in a Gibbs ensemble.

If in this totality of systems

$$\overline{\Delta p^2} \cdot \overline{\Delta x^2} \cong \frac{\hbar^2}{4}, \tag{2}$$

we call the set a quantum ensemble.[2,3,4]

Figure 4 illustrates a quantum ensemble as an indefinitely receding row of macroscopic bodies $\mathcal{M}$ and microsystems μ.

The macroscopic setting is constituted not only by macroscopic bodies but also by macroscopic fields, and it includes the conditions of generation of the particles (particle sources).

The more exact definition of a quantum ensemble contains no assumption about absolute-zero temperature, by which previously we wished to stress that the thermal motion of atoms or molecules has no effect on the statistics of quantum effects. In fact, quantum effects are also observed in this case when the temperature of the bodies is above zero. Here we speak of quantum statistics, which acts as a correction to the Gibbs distribution at low temperatures. More precisely, we may say that we are concerned with quantum mechanics if (2) is obeyed and if the kinetic energy of the particles $T(p) = p^2/2m \gg \theta$, with m being the mass of a particle, whereas we are concerned with quantum statistics if $T(p) \approx \theta$. We now must examine whether we can describe a quantum ensemble via a probability of the type constituting the basic law of a Gibbs ensemble. *A priori* there would appear to be no reason why this should not be possible.

In fact, this is not the case. Relation (1) states that this possibility is

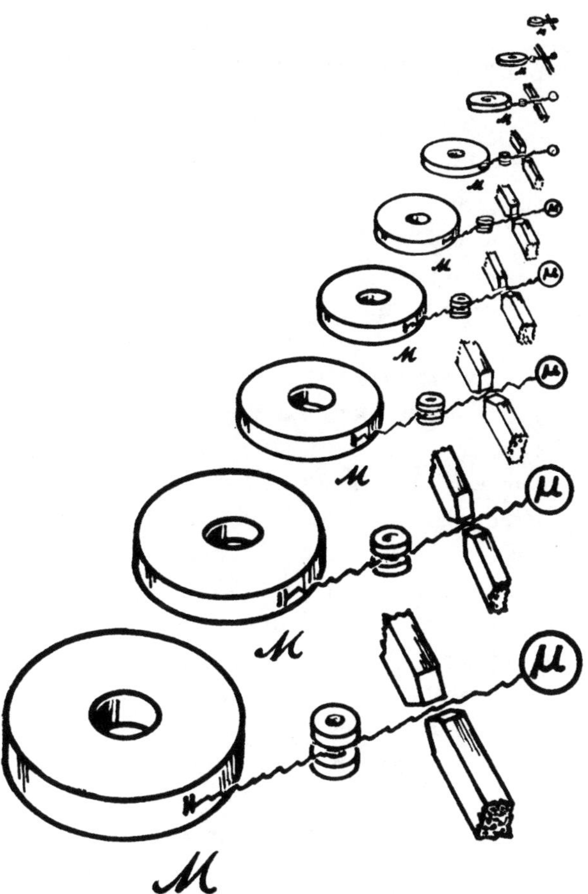

Fig. 4. A quantum ensemble: an infinite sequence of identical microsystems μ each in the same macroscopic setting $\mathcal{M}$ (e.g. an accelerator operating in a given mode, a magnetic analyser, collimating slit, etc.).

ruled out, for it indicates that there is no quantum ensemble for which $\overline{\Delta q^2}$ and $\overline{\Delta p^2}$ are simultaneously zero; if there were a probability $W(p, q)$ that would predict the possibility of finding a microsystem with q as its coordinate and p as its momentum, then it would be possible to state a method of selecting such systems and hence to construct from these a new ensemble having $\overline{\Delta p^2}=0$ and $\overline{\Delta q^2}=0$. Such an ensemble would conflict with the uncertainty principle.

We must thus conclude that there is no such probability $W(p, q)$, and we must examine what replaces it for quantum ensembles, how the latter may be characterized, and how the probability of different results of observation may be calculated.

The problem may be dealt with by formulating Heisenberg's relation in somewhat more general form, via Bohr's principle of complementarity, which can be put as follows:

The dynamic variables of a microsystem may be divided into two mutually complementary groups: space-time ones and momentum-energy ones. There is no ensemble in which both groups of dynamic variables could have definite values.

This is a direct generalization of Heisenberg's relation. Bohr himself formulated the principle in a somewhat different way, reflecting his philosophical concepts which were far from those of materialism. His formulation has been the origin of the far-reaching conclusion that the current mechanics of the atom cannot be compatible with materialism. We shall not deal in any further detail with this aspect of the question, but there is an extensive literature[5,6,7] presenting various points of view.

It would seem generally better to speak of a principle of exclusiveness rather than complementarity: dynamic variables should be divided into mutually exclusive groups, which do not coexist in real ensembles. However, out of respect for Bohr and his tradition we shall retain the usual terminology.

Complementarity rules out the description of microsystems via a phase space $R(p, q)$, since this contains the complementary variables p and q; however, the principle does not forbid us to use either configuration space $R(q)$ or momentum space $R(p)$, as each of these contains variables that are either only spatial or only momental. A system with f degrees of freedom is described by f variables either in $R(q)$ or in $R(p)$, in contrast to classical mechanics. The variables p and q form a complete set of variables.

More precisely, we define the complete set of dynamic variables sufficient for an exhaustive description of a quantum system with f degrees of freedom as a set of f simultaneously measurable independent variables: $q_1, q_2, ..., q_f$ (spatial, set of q) or $p_1, p_2, ..., p_f$ (momentum, set of p). Such a set of dynamic variables contains the maximum information on the system compatible with the laws that govern the microworld.

A quantum ensemble therefore has the probability

$$W_M(q)\,\mathrm{d}q \tag{3}$$

that the specific value q will be found in a measurement. The subscript M has the same meaning as the subscript θ (temperature) in Equation (III–1), for a classical Gibbs ensemble; it indicates the macroscopic setting (M) that governs the conditions of motion for the microsystem (μ) and hence determines its state.

A further question is the probability that a microsystem in the same ensemble will have some particular value in the complementary (p) set. This probability may be put as

$$W_M(p)\,\mathrm{d}p. \tag{4}$$

The number of probabilities of the type of (3) or (4) is not necessarily small; in principle, the number may be as large as the number of possible distinct complete sets (q) or conjugate sets (p).

The number for any particular real system is restricted by the fact that many of the sets are not convenient for description of the system, but this is a purely practical consideration, so the number of probabilities such as (3) or (4) may be indefinitely large. The set of these characterizes completely the state of a microsystem in a quantum ensemble, since this set essentially exhausts the predictions of all possible measurements on a microsystem in this ensemble $(M+\mu)$.

This raises the problem of counting the probabilities, including those unsuitable for practical purposes. If a way of calculating the number can be found, we will have a method of describing the state of a microsystem in a quantum ensemble.

The answer given by quantum mechanics to this question is quite unexpected (from the viewpoint of the classical theory): there is a quantity characteristic of the ensemble $\mathcal{M}+\mu$ that characterizes the quantum ensemble completely in the sense that, when the quantity is known, one can calculate all possible probabilities of the type $W_{\mathcal{M}}(q)$ or $W_{\mathcal{M}}(p)$. This quantity is the wave function:

$$\psi_{\mathcal{M}} = \psi_{\mathcal{M}}(q). \tag{5}$$

This $\psi_{\mathcal{M}}$ is written explicitly as a function of the coordinates q, the relation of the wave function to the probability density in this case being given by

$$W_{\mathcal{M}}(q) = |\psi_{\mathcal{M}}(q)|^2 . \tag{6}$$

The above formula is completely general in the following sense: the probability density for the value p in the set of variables (p) is as follows:

$$W_{\mathcal{M}}(p) = |\psi_{\mathcal{M}}(p)|^2 . \tag{6'}$$

Here the wave function $\psi_{\mathcal{M}}$ is given as a function of the variables of some other complete set, for example $\psi_{\mathcal{M}}(p)$. More generally, for a complete set of variables (a) we have

$$W_{\mathcal{M}}(a) = |\psi_{\mathcal{M}}(a)|^2 , \tag{7}$$

and for a set (b)

$$W_{\mathcal{M}}(b) = |\psi_{\mathcal{M}}(b)|^2 \tag{7'}$$

etc. All these functions $\psi_{\mathcal{M}}(q)$, $\psi_{\mathcal{M}}(p)$, $\psi_{\mathcal{M}}(a)$, $\psi_{\mathcal{M}}(b)$, and so on describe the same quantum ensemble, which is characterized by a macrosetting $\mathcal{M}$ and a microsystem μ.

This feature is stressed by saying that the wave function is given in the q-representation or in the p-representation, in the a-representation, and so on.

The mathematical equivalence of the wave functions in the various representations, as regards description of the quantum ensemble, is seen from the fact that $\psi_{\mathcal{M}}$ may be considered as a vector in Hilbert space, while the representations may be considered as representations of the vector in different systems of loci (orthogonal unit vectors), a linear combination of these allowing one to represent any other vector.[2]

In this formulation, rotation in Hilbert space represents a transition from one representation of the wave function to another. This rotation is performed via a unitary matrix S, so that

$$\psi_{\mathcal{M}}(b) = S\psi_{\mathcal{M}}(a), \tag{8}$$

with

$$SS^+ = 1 , \tag{9}$$

in which S^+ is the matrix conjugate with S. Relations (8) and (9) in explicit form state that

$$\psi_{\mathcal{M}}(b) = \int S(b/a)\,\psi_{\mathcal{M}}(a)\,\mathrm{d}a , \tag{8'}$$

$$\int S(b/c)\, S^+(c/a)\, \mathrm{d}c = \delta(b - a), \qquad (9')$$

in which the integrals in (8') and (9') are to be understood in the generalized sense, as being suitable for cases where the spectra of possible values for a and b may have discontinuities or may even consist of separate points (discrete spectra). In the latter case the integral reduces to a sum. The matrices S are derived from linear equations (see textbooks on quantum mechanics, e.g.[4]).

The transform of (8) allows us to consider the wave function as an objective characteristic of a quantum ensemble, but one which may be given in various representations. This important feature forms the basis of the concept of a quantum ensemble. In this concept the wave function $\psi_{\mathcal{M}}$ is considered as the equivalent, in quantum theory, of the classical probability $W_\theta(p, q)$ of a state for the system in phase space $R(p, q)$. A knowledge of $W_\theta(p, q)$ enables us to determine the probability of any other set (P, Q) of dynamic variables in phase space; similarly, a knowledge of $\psi_{\mathcal{M}}$ for a quantum ensemble enables us to determine the probability of any complete set (a) of dynamic variables. Hence the wave function is not the quantity that determines the statistics of any particular measurement but one that determines the statistics of a quantum ensemble, i.e. the statistics of any measurement compatible with the nature of the microsystem μ and of the macroscopic setting $\mathcal{M}$ that governs the conditions of motion for μ.

REFERENCES

1. R. W. James, G. W. Brindley, and R. J. Wood, *Proc. Roy. Soc.* A **125** (1929) 401.
2. J. von Neumann, *The Mathematical Basis of Quantum Mechanics* [Russian translation], Nauka, 1964.
3. K. V. Nikol'skii, *Quantum Processes* [in Russian], GITTL, 1940.
4. D. I. Blokhintsev, *The Principles of Quantum Mechanics* [in Russian]. Izd. Vysshaya Shkola, 4th ed., 1963.
5. N. Bohr, *Uspekhi fiz. nauk* **64** (1958) 571.
   N. Bohr, *Library of Living Philosophers: A. Einstein*, 1949, p. 201.
6. V. A. Fok, *Uspekhi fiz. nauk* **62** (1957) 461.
7. D. I. Blokhintsev, *ibid.* **55** (1956) 195.

CHAPTER V

THE DENSITY MATRIX

In certain cases the macroscopic setting is not sufficiently defined and must itself be given by a statistical description. This naturally introduces a further uncertainty into the quantum ensemble, which is then called *mixed*, to distinguish it from the *pure* ensemble that occurs when the macroscopic setting is completely defined (see Chapter IV and also refs. 2 and 4).

A very simple example of this situation occurs when there are several incoherent sources of particles, which may themselves differ from one another in momentum, polarization, or other parameters. Let the first source generate a particle with a probability $\mathscr{P}_{\mathscr{M}_1}$ and the second one with a probability $\mathscr{P}_{\mathscr{M}_2}$ ($\mathscr{M}_1$ and $\mathscr{M}_2$ denote the macroscopic settings of the sources). The first source could produce a pure quantum ensemble described by the wave function $\psi_{\mathscr{M}_1}(q)$, the second similarly producing one described by $\psi_{\mathscr{M}_2}(q)$. We thus have a quantum ensemble that must be described via a set of probabilities:

$$\mathscr{P}_{\mathscr{M}_1} \quad \text{and} \quad \mathscr{P}_{\mathscr{M}_2}$$

and a set of wave functions

$$\psi_{\mathscr{M}_1}(q) \quad \text{and} \quad \psi_{\mathscr{M}_2}(q).$$

With $\mathscr{P}_{\mathscr{M}_1} + \mathscr{P}_{\mathscr{M}_2} = 1$, we have $\mathscr{P}_{\mathscr{M}_1}$ and $\mathscr{P}_{\mathscr{M}_2}$ as statements of the proportions in which the pure ensembles are mixed, these ensembles being described by $\psi_{\mathscr{M}_1}$ and $\psi_{\mathscr{M}_2}$ in our mixed ensemble ($\mathscr{M} = \mathscr{M}_1 + \mathscr{M}_2$). It is clear that in the general case we have any set of probabilities (including an indefinitely large one):

$$\mathscr{P}_{\mathscr{M}_1}, \mathscr{P}_{\mathscr{M}_2}, ..., \mathscr{P}_{\mathscr{M}_s}, ..., \quad \sum_s \mathscr{P}_{\mathscr{M}_s} = 1 \tag{1}$$

and the corresponding wave functions

$$\psi_{\mathscr{M}_1}, \psi_{\mathscr{M}_2}, ..., \psi_{\mathscr{M}_s}, ... \tag{2}$$

This description via the two series of $\mathscr{P}_{\mathscr{M}_s}$ and $\psi_{\mathscr{M}_s}$ is very inconvenient and, at first sight, seems to take us far from the analogy with the classical Gibbs ensemble. However, the analogy is restored if we replace the wave function in the description of a mixed quantum ensemble by a quadratic form derived from the wave function, which is termed the density matrix:

$$\rho_{\mathscr{M}}(q, q') = \sum_{\mathscr{M}_s} \mathscr{P}_{\mathscr{M}_s} \cdot \psi^*_{\mathscr{M}_s}(q') \, \psi_{\mathscr{M}_s}(q); \tag{3}$$

in which q and q' denote two different points in the space of some complete set of dynamic variables (configuration or momentum ones). The q points correspond to the horizontal lines of the matrix, and the q' points to the columns. A diagonal element $(q=q')$ represents the probability of finding the value q for the dynamic variables in a mixed quantum ensemble. In fact, in this case we get the usual formula for combining the probabilities of independent events:

$$W_{\mathscr{M}}(q) = \rho_{\mathscr{M}}(q, q) = \sum_{\mathscr{M}_s} \mathscr{P}_{\mathscr{M}_s} |\psi_{\mathscr{M}_s}(q)|^2 . \tag{4}$$

For the particular case of $\mathscr{P}_{\mathscr{M}_s} = 1$, all other $\mathscr{P}_{\mathscr{M}_s}$ are zero, and we return to the initial formula for a pure ensemble:

$$W_{\mathscr{M}}(q) = |\psi_{\mathscr{M}}(q)|^2 . \tag{4'}$$

The elements of this density matrix satisfy certain symmetry conditions. $\psi_{\mathscr{M}}(q)$ and its conjugate $\psi^*_{\mathscr{M}}(q)$ describe the same state; (3) implies that $\rho(q, q')$ is the Hermite matrix:

$$\rho_{\mathscr{M}}(q', q) = \rho^*_{\mathscr{M}}(q, q') = \rho^+_{\mathscr{M}}(q', q). \tag{5}$$

Further, if the microsystem itself consists of identical particles (or does so at least partly), then $\psi_{\mathscr{M}}(q)$ either remains unchanged when identical particles i and k are interchanged (particles obeying Bose-Einstein statistics) or reverses its sign (particles obeying Fermi statistics). Let $\mathscr{P}_q$ be the operator denoting interchange of the dynamic variables q_i and q_k of particles i and k; we then have for the wave function

$$\mathscr{P}_q \psi_{\mathscr{M}}(q) = \pm \, \psi_{\mathscr{M}}(q). \tag{6}$$

The density matrices are then

$$\mathscr{P}_q \rho_{\mathscr{M}}(q', q) = \pm \, \rho_{\mathscr{M}}(q', q). \tag{7}$$

27

$$\mathscr{P}_{q'}\rho_{\mathscr{M}}(q', q) = \pm \, \rho_{\mathscr{M}}(q', q). \qquad (7')$$

$$\mathscr{P}_{qq'}\rho_{\mathscr{M}}(q', q) = + \, \rho_{\mathscr{M}}(q', q). \qquad (8)$$

The last relation indicates that the matrix is symmetric with respect to particle interchange ($\mathscr{P}_{qq'} = \mathscr{P}_q \mathscr{P}_{q'}$ in operators).

Let some function of the dynamic variables be represented by the operator $\mathscr{L}$, whose matrix elements in the q representation are $L(q', q)$; quantum mechanics shows that the mean value L of the physical quantity represented by $\mathscr{L}$ is

$$\bar{L} = \int \psi_{\mathscr{M}}^{*}(q') \, L(q', q) \, \psi_{\mathscr{M}}(q) \, dq' \, dq \, . \qquad (9)$$

Then (3) shows that this mean in a mixed ensemble described by the matrix of (5) is

$$\bar{L} = \int \rho_{\mathscr{M}}(q, q') \, L(q', q) \, dq' \, dq = S_p(\rho_{\mathscr{M}}\mathscr{L}), \qquad (10)$$

in which S_p denotes the spur of matrix $(\rho_{\mathscr{M}}\mathscr{L})$.

The density matrix may also be given in a mixed representation, the lines denoting the values of one set of dynamic variables (q) and the columns those of another set (p):

$$\rho_{\mathscr{M}}(q, p) = \sum_{\mathscr{M}_s} \mathscr{P}_{\mathscr{M}_s} \cdot \psi_{\mathscr{M}_s}^{*}(q) \cdot \psi_{\mathscr{M}_s}(p), \qquad (11)$$

whereupon, in accordance with the general rules for transforming the wave function from one set of variables to another, we have:

$$\psi_{\mathscr{M}_s}(p) = \int S(p, q) \, \psi_{\mathscr{M}_s}(q) \, dq, \qquad (12)$$

in which S is the matrix for the unitary transformation from the q variables to the p variables.

If by the q variables we understand the coordinates and by the p ones the momenta, then the density matrix becomes the direct analogue of the classical density in phase space, $W(q, p)$. However, there is[1] a more direct relation between the classical density in phase space and the quantity

$$R_{\mathscr{M}}(q, p) = \rho_{\mathscr{M}}(q, p) \, S^{-1}(q, p), \qquad (13)$$

in which $S^{-1}(q, p)$ is an element of the matrix inverse to $S(q, p)$. The matrix S has the following elements when the coordinates q and momenta p are cartesian:

$$S(q, p) = \frac{\exp((ipq/\hbar))}{\sqrt{2\pi\hbar}} \tag{14}$$

and all the relationships become particularly simple. It is readily shown from (11), (13) and (14) that

$$R_{\mathcal{M}}(q) = \int R_{\mathcal{M}}(q, p)\, \frac{\mathrm{d}p}{2\pi\hbar} \tag{15}$$

is the probability of finding a value q for the coordinates in the ensemble, while

$$R_{\mathcal{M}}(p) = \int R_{\mathcal{M}}(q, p)\, \frac{\mathrm{d}q}{2\pi\hbar} \tag{16}$$

is the probability of finding momenta equal to p. Formulae (15) and (16) coincide exactly with those of classical theory if $R_{\mathcal{M}}(q, p)$ is considered as $R(q, p)$, the density in phase space. Finally, we have the formula for the mean $\bar{L}$ of the physical quantity represented by the operator $\mathcal{L}$ whose matrix elements are $L(q, p)$ in the $q-p$ representation:

$$\bar{L} = \int R_{\mathcal{M}}^{*}(q, p)\, L(q, p)\, \frac{\mathrm{d}q\, \mathrm{d}p}{2\pi\hbar}, \tag{17}$$

in which $R_{\mathcal{M}}^{*}(q, p)$ is the complex conjugate of matrix $R_M(q, p)$, while

$$L(q, p) = \int L(q, q')\, S(q', p)\, \mathrm{d}q' \cdot S^{-1}(q, p). \tag{18}$$

Formulae (15)–(17) are so closely analogous to the corresponding formulae of classical mechanics that one is tempted to base the description of a quantum ensemble on the density matrix $R(q, p)$ rather than on the wave function; however, a more detailed examination shows that this temptation should be resisted. This aspect will be discussed in more detail later; here we merely note that rather cumbrous conditions replace the simple ones for the wave function and the matrix when the arguments (dynamic variables of identical particles) are permuted, as do the conditions for hermiticity in the operators that represent physical quantities in the language of the $R(q, p)$ matrix. For instance, it is readily shown[1] that (5) leads to the integral relation

$$R^*(q, p) = \int R(q + \xi, p + \eta) \exp\left(i\,\frac{\xi\eta}{\hbar}\right) \cdot \frac{d\xi\,d\eta}{2\pi\hbar}, \tag{19}$$

while conditions (7) and (7′) (for symmetry in the interchange of identical particles, become

$$\mathscr{P}_q R(q, p) = \pm\, R(q, p) \exp\left(-\frac{i}{\hbar}(p_i - p_k)(q_i - q_k)\right), \tag{20}$$

$$\mathscr{P}_p R(q, p) = \pm\, R(q, p) \exp\left(-\frac{i}{\hbar}(p_i - p_k)(q_i - q_k)\right), \tag{20′}$$

$$\mathscr{P}_{pq} R(q, p) = R(q, p), \quad \mathscr{P}_{pq} = \mathscr{P}_p \mathscr{P}_q, \tag{20″}$$

in which q_i and q_k are the coordinates, and p_i and p_k are the momenta of the interchanged particles i and k. $\mathscr{P}_q$ is the coordinate interchange operator and $\mathscr{P}_p$ the momentum interchange operator.

All these relationships are readily satisfied if $R(q, p)$ is considered as a bilinear form from the wave function given in the $q-p$ representation.

The practical value of the density matrix $R(q, p)$ is that it may be expanded as a power series in Planck's constant if the quantum ensemble differs little from a classical one:

$$R = \sum_{n=0}^{\infty} \hbar^n \cdot R_n. \tag{21}$$

Condition (19) may also be put as a power series in $\hbar$, for which purpose we expand $R(q+\xi, p+\eta)$ in powers of ξ and η, then using the equation

$$I_{nm} = \int\int \xi^n \cdot \eta^m \cdot \exp\left(\pm\frac{i\xi\eta}{\hbar}\right) \cdot \frac{d\xi\,d\eta}{2\pi\hbar} = (\pm\, i\hbar)^{n+m}\frac{}{m!}\,\delta_{nm}. \tag{22}$$

This gives us from (19) that

$$R^*(q, p) = \sum_{n=0}^{\infty} \frac{(i\hbar)^n}{n!}\,\frac{\partial^{2n} R(q, p)}{\partial q^n\,\partial p^n} \tag{23}$$

and, substituting from (21), we get

$$R_0 = R_0^*, \; R_1^* = R_1 + \frac{i}{1!} \frac{\partial^2 R_0}{\partial q \, \partial p},$$

$$R_2^* = R_2 + \frac{i}{1!} \frac{\partial^2 R_1}{\partial q \, \partial p} - \frac{1}{2!} \frac{\partial^4 R_0}{\partial q^2 \, \partial p^2}. \tag{24}$$

These relationships show that matrix $R(q, p)$ can be real only in the limit $\hbar \to 0$, i.e. for a classical ensemble; it is obliged to have a non-zero imaginary part for $\hbar \neq 0$, so the matrix $R(q, p)$ is not the probability density in phase space $R(q, p)$, as we would expect from the uncertainty principle, which forbids an ensemble with exactly known coordinates q and momenta p.

A special position arises for an ensemble containing identical particles. The symmetry conditions (20) and (20′) contain Planck's constant in a form such that a singularity is bound to arise for $\hbar \to 0$; this makes it impossible to expand R as a power series in $\hbar$. This difficulty is avoided[2,3] if the matrix R is replaced by a matrix averaged over the phase volume:

$$\tilde{R} = \frac{1}{\Omega} \int_\Omega R \, \mathrm{d}p \, \mathrm{d}q, \tag{25}$$

in which Ω is the volume of the region used in the average:

$$\mathscr{P} - \varDelta < p < \mathscr{P} + \varDelta, \quad Q - \delta < q < Q + \delta, \quad \varDelta \cdot \delta = \Omega, \tag{25'}$$

The significance of this average is as follows. Relations (20) and (20′) imply that R in Fermi statistics becomes zero on all hyper-planes in phase space for which $p_i = p_k$ or $q_i = q_k$ (neglecting complications which arise from the spin), while for Bose-Einstein statistics the matrix has turning points with respect to the variable $p_i - p_k$ or $q_i - q_k$ on the surfaces $p_i = p_k$, $q_i = q_k$ (matrix ρ, which may be considered as the amplitude of matrix R, is seen from (13) to have turning points on the surfaces $p_i = p_k$ and $q_i = q_k$). A region of the zero value or turning point extends over roughly the wavelength $\lambdabar \simeq h/p$ of the particle. These regions do not play an important part if there is only a small probability of finding two particles in a distance comparable with the wavelength; this probability is $\lambdabar^3 n$ if there are n particles per unit volume, and for $\lambdabar^3 n \ll 1$ we can choose a volume Ω such that there is only a small contribution from the anomalous regions (where $q_i \approx q_k$ and $p_i \approx p_k$) in the whole of phase space.

Condition $\lambda^3 n \ll 1$ may be put as $p^3 n \gg \hbar^3$, which means that the phase volume per particle must be large relative to λ^3; if this is so, we can choose a volume Ω for averaging R, such that $p^3/n \gg \Omega \gg \hbar^3$. The $\tilde{R}$ of (25) then has[2,3] no important singularities with respect to $\hbar$.

REFERENCES

1. P. A. M. Dirac, *The Principles of Quantum Mechanics*, section 33, 3rd ed., 1947.
2. D. I. Blokhintsev, *Journ. of Phys. USSR* **2** (1940) 71.
3. D. I. Blokhintsev and P. Nemirowsky, *Journ. of Phys. USSR* **3** (1940) 191; *ZhEFT* (*JETP*) **10** (1940) 1263.

CHAPTER VI

CAUSALITY IN QUANTUM MECHANICS

Quantum mechanics is essentially a statistical theory, so that the history of an individual particle may be traced only in very general terms (see Chapter XV). We have seen that even in classical mechanics the input data must be corrected from time to time to correct cumulative effects arising from errors in the initial values. This feature becomes extremely important in quantum mechanics: a particle may be localized in space only approximately, and the localization steadily deteriorates as time passes, and the more rapidly the more precise was the original localization; this is a direct consequence of the uncertainty principle.

This is in no way an attempt to belittle the value of quantum mechanics. All the more so, the statistical character of quantum mechanics should not be understood (as it is sometimes done) as indicating incompleteness or the need to seek a completely deterministic theory, for there may be no such theory; of course, we should not interfere with anyone who is attempting the apparently impossible.

From what has been shown above regarding the illusion of determinism in the classical theory, where at least it had a certain basis in that the initial data could be discarded relatively slowly, it is natural to conclude that this illusion becomes for the microworld simply dangerous self-deception. However, this does not mean that causality has no place in statistical theories.

Causality is a definite form of order in events in space and time; this ordering imposes restrictions even on the most chaotic events, and it makes itself felt in two ways in statistical theories. Firstly, the statistical laws themselves are fully ordered, and the quantities that characterize an ensemble are themselves strictly determined. Secondly, the individual elementary events are also so ordered that one may influence another only if their relative location in space and time allows this without violating causality (i.e. the rule ordering the events).

In the relativistic theory, the events must be linked by a light signal or by some other signal propagating with a velocity lower than that of light. In the non-relativistic theory (the one envisaged here) all the velocities are so much slower than that of light, that the latter may be taken as infinitely high, the theory being constructed formally in a way such as to allow signals that produce interaction to travel with infinitely high speeds. Figure 5 shows the "spheres of influence" in the relativistic and non-relativistic cases. In the latter, each point on the plane $t=0$ influences the points on an infinitely close plane $t+\mathrm{d}t$, because the light cone opens into a plane for $c \rightarrow \infty$. Determinism in the classical case then means simply that the state of the system at a preceding instant completely determines the state at a subsequent instant.

The state of a system means in quantum theory that the system belongs to some quantum ensemble, which is characterized by a wave function

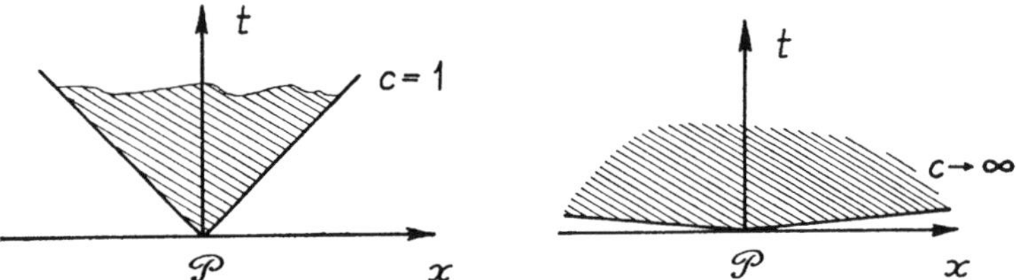

Fig. 5.　Region within which an event at point P can influence events at $t > 0$:
a) relativistic theory, b) classical theory.

in the pure case, or by a density matrix in the general case. Consider first the pure case. Since the wave function Ψ as a time function provides an exhaustive description of the state of the ensemble, the above formulation of the causality concept implies that the subsequent state of the ensemble is also determined by the wave function. This Ψ at time t has an increment $\mathrm{d}\Psi$ in time $\mathrm{d}t$, which must itself be expressed in terms of Ψ. Further, this wave function itself is essentially defined with an arbitrary factor N, so Ψ and $\Psi'' = N\Psi$ represent the same state. This factor must have a modulus of 1 for the normalized function (i.e. $N = \exp(i\alpha)$ in which α is an arbitrary real number), so $\mathrm{d}\Psi$ must be linearly related to Ψ. We have

$$\mathrm{d}\Psi = \mathscr{L}\Psi, \tag{1}$$

in which $\mathscr{L}$ is some linear operator. Let the wave function at time $t+\mathrm{d}t$ be $\Psi' = \Psi + \mathrm{d}\Psi$; then the previous relation may be put as

$$\Psi' = S(\mathrm{d}t)\,\Psi,$$

in which $S(\mathrm{d}t)$ denotes a matrix that transforms Ψ into Ψ'. The conservation of

$$\int |\Psi|^2 \, \mathrm{d}q$$

implies that this matrix is unitary, i.e. that

$$S^+ (\mathrm{d}t) \equiv \tilde{S}^* (\mathrm{d}t) = S^{-1} (\mathrm{d}t).$$

We put $S = 1 + \mathscr{L} \, \mathrm{d}t + \cdots$ to get

$$\mathscr{L} = - (i/\hbar) \mathscr{H}$$

in which the operator $\mathscr{H}$ is hermitian ($\mathscr{H}^* = \mathscr{H}^+ = \mathscr{H}$), while $\hbar$ is introduced as a factor from dimensional considerations and correspondence with the classical theory. $\mathscr{H}$ is essentially the operator of displacement in time and is Hamilton's operator. If $\mathscr{H}$ itself is not dependent on time, it is simply the energy operator of the microsystem.

This shows that the equation for the variation of the wave function in time must be put in the form

$$i\hbar \, \frac{\partial \Psi}{\partial t} = \mathscr{H} \Psi. \tag{2}$$

This is Schrödinger's equation, which expresses causality in quantum theory; Ψ describes the quantum ensemble completely, so we may say that (2) describes the motion of the quantum ensemble and that in a causal fashion, i.e. so that the state earlier in time determines the subsequent state of the ensemble.

The equation for the conjugate function Ψ^* states that

$$- i\hbar \, \frac{\partial \Psi^*}{\partial t} = \mathscr{H}^* \Psi^*, \tag{2'}$$

and for $\mathscr{H}^* = \mathscr{H}$ we have

$$- i\hbar \, \frac{\partial \Psi^*}{\partial t} = \mathscr{H} \Psi^*. \tag{2''}$$

The differential equation of (2) may be replaced by the corresponding integral equation. Green's functions may be used for a system of non-interacting particles. Essentially, $\mathscr{H}$ may usually be put as the sum of the operator T for the kinetic energy of the particles and V, the operator for the interactions of the particles mutually and with external fields. T has a particularly simple form in a cartesian coordinate system:

$$T = -\sum_{s=1}^{N} \frac{\hbar^2}{2m_s}\left(\frac{\partial^2}{\partial x_s^2} + \frac{\partial^2}{\partial y_s^2} + \frac{\partial^2}{\partial z_s^2}\right), \tag{3}$$

in which m_s are masses of particles $(s=1, 2, ..., N$, with N the number of particles in the system), whose cartesian coordinates are x_s, y_s, z_s. The Green's function G by definition is the solution to a Schrödinger equation whose right-hand side takes the form of an instantaneous source at point $x=x'$ and time $t=t'$

$$i\hbar\,\frac{\partial G}{\partial t} - TG = -\,\delta(t - t')\,\delta(x - x') \tag{4}$$

being zero for $t<t'$. (Here by x and x' are meant all the coordinates of the particles, so that $\delta(x-x')$ is a δ-function in $3N$ dimensions).

This equation coincides with the diffusion equation in multi-dimensional space, but with an imaginary diffusion coefficient and an instantaneous particle source at $x=x'$, $t=t'$. The solution to the equation is

$$G(x - x', t - t') = \prod_{s=1}^{N} \frac{1}{[i\hbar(t - t')]^{N/2}}\cdot\exp\left(-\,m_s\frac{(x_s - x'_s)^2}{4i\hbar(t - t')}\right) \tag{5}$$

for $t'<t$ and

$$G(x - x', t - t') = 0. \tag{5'}$$

for $t'>t$.

We may put (2) in the form

$$i\hbar\,\frac{\partial \Psi(x', t')}{\partial t'} - T(x')\,\psi(x', t') = V(x', t')\,\Psi(x', t'). \tag{2'''}$$

Multiplying (4) by $V(x', t')\,\Psi(x', t')$ and integrating with respect to x' and t' we find that $\Psi(x, t)$ satisfies the integral equation

$$\Psi(x, t) = \Psi_0(x, t) + \int G(x - x', t - t')\,V(x', t')\,\Psi(x', t')\,\mathrm{d}x'\,\mathrm{d}t', \tag{6}$$

in which $\Psi_0(x, t)$ is the solution for $V=0$.

This equation shows that there will be no effect on the state of the ensemble, i.e. on $\Psi(x, t)$, from a change $\delta\Psi(x', t')$ in the wave function or an interaction $\delta V(x', t')$ occurring at point x' at time t' (an event) if $t' > t$; in other words, the state at any instant t is influenced only by events occurring at $t' < t$.

Equation (2) not only gives us the state in the immediate future from the present state, but also the state in the immediate past; we merely have to take $dt < 0$. The "delayed" Green's function of (5) may thus be considered together with an "advanced" one:

$$G(x - x', t - t') = \prod_{s=1}^{N} \frac{1}{[-i\hbar(t - t')]^{N/2}} \exp\left(+ m_s \frac{(x_s - x'_s)^2_s}{4i\hbar(t - t')} \right) \quad (5'')$$

for $t' > t$ and

$$G(x - x', t - t') = 0 \quad (5''')$$

for $t' < t$. This function enables us to determine the state in the past. This possibility arises from the reversibility of quantum mechanics in time.

A. EQUATIONS OF MOTION FOR A MIXED ENSEMBLE, DENSITY MATRIX $\rho(q, q')$

We differentiate Equation (V–3) with respect to time on the assumption that the conditions determining the mixture remain unchanged, so that $P_s =$ constant; we obtain

$$\frac{\partial \rho(q, q')}{\partial t} = \sum_s \mathscr{P}_{M_s} \left\{ \frac{\partial \Psi^*_{M_s}(q')}{\partial t} \Psi_{M_s}(q) + \Psi^*_{M_s}(q') \frac{\partial \Psi_{M_s}(q)}{\partial t} \right\} \quad (7)$$

and use (2) and (2′) to get that

$$\frac{\partial \rho(q, q'')}{\partial t} = \sum_s \mathscr{P}_{M_s} \frac{1}{i\hbar} \left\{ - \mathscr{H}^* \Psi^*_{M_s} \cdot \Psi_{M_s}(q) + \Psi^*_{M_s}(q') \cdot \mathscr{H} \Psi_{M_s}(q) \right\}, \quad (7')$$

or, since

$$\mathscr{H}(q', q'') = \widetilde{\mathscr{H}}^*(q', q'') = \mathscr{H}^+(q', q''),$$

we have

$$\frac{\partial p(q, q')}{\partial t} + [\mathscr{H}, \rho] = 0, \quad (8)$$

in which $[\mathscr{H}, \rho]$ is the quantum Poisson bracket:

$$[A, B] = \frac{1}{i\hbar}(AB - BA). \tag{9}$$

Equation (8) may be put in the form

$$\frac{d\rho}{dt} = 0, \tag{10}$$

which expresses Louiville's theorem in quantum mechanics; (8) or (10) expresses a causality essentially equivalent to that expressed by (2): the first derivative of the density matrix ρ is itself defined by that matrix, i.e. the future state is determined by the previous state.

If from $\rho(q, q')$ we pass to the matrix $R(q, p)$, we find (see references[1,2,3] of Chapter V) that the following expression for the matrix R after simple manipulations is:

$$\frac{\partial R(q, p)}{\partial t} + [\mathscr{H}, R] \equiv \frac{dR}{dt} = 0, \tag{11}$$

where

$$[\mathscr{H}, R] = \frac{1}{i\hbar} \int \frac{d\xi \, d\eta}{2\pi\hbar} \exp\left(-\frac{i\xi\eta}{\hbar}\right)$$

$$\{H(q, p + \eta) R(q + \xi, p) - H(q + \xi, p) R(q, p + \eta)\}. \tag{11'}$$

We expand this integral as a power series in $\hbar$, for which purpose we use formula (V–22):

$$\frac{\partial R(q, p)}{\partial t} + \sum_{n=1}^{\infty} \frac{(-i\hbar)^{n-1}}{n!} \cdot [H, R]_n = 0, \tag{12}$$

in which

$$[H, R]_n = \frac{\partial H}{\partial p^n} \frac{\partial R}{\partial q^n} - \frac{\partial H}{\partial q^n} \frac{\partial R}{\partial p^n} \tag{13}$$

in the classical Poisson bracket of order η. If $\mathscr{H}$ equals $T(p) + V(q)$, the matrix elements of $\mathscr{H}$ in the q, p-representation are simply equal to the corresponding classical hamiltonian:

$$H(q, p) = \frac{p^2}{2m} + V(q). \tag{14}$$

38

Then (12) takes the form

$$\frac{\partial R}{\partial t} = -\frac{p}{m}\frac{\partial R}{\partial q} + \frac{\partial V}{\partial q}\frac{\partial R}{\partial p} + \frac{i\hbar}{2m}\frac{\partial^2 R}{\partial q^2} - \sum_{n \geq 2}^{\infty} \frac{(-i\hbar)^{n-1}}{n!}\frac{\partial^n V}{\partial q^n}\frac{\partial^n R}{\partial p^n}. \qquad (15)$$

The first two terms in this equation coincide exactly with the classical equation for the density $\rho(q, p)$ [see (IV–4) or (IV–5)], while the subsequent terms may be considered as quantum corrections, which are dependent on the higher derivatives of the density R and of the potential V, which include derivatives with respect to p as well as ones with respect to q. These equations may thus be useful when the density function R is a reasonably smooth function in the phase space $R(p, q)$.

The first two (classical) terms in (15) show that the motion in the classical approximation is governed by the speed p/m at the point in question and by the force $\partial V/\partial q$ acting at that point. The subsequent (quantum) terms indicate the steadily increasing importance of the environment in (p, q) phase space: the $\partial^2 R/\partial q^2$ term leads to diffusion of the density $R(q, p)$ (diffusion coefficient imaginary) while terms of the type $(\partial^n V/\partial q^n)(\partial^n R/\partial p^n)$ indicate that the motion of the ensemble in quantum mechanics is governed not only by the form of the potential at the point in question but also by that throughout the space accessible to the ensemble.

B. EXPLICIT FORM OF THE EQUATION FOR THE DENSITY MATRIX $\rho(q, q')$

This is found in the coordinate representation by noting that the matrix elements of $\mathcal{H}$ take the following form in the q representation:

$$H(q, q') = +\frac{\hbar^2}{2m}\frac{\partial^2}{\partial q\,\partial q'}\,\delta(q - q') + V(q')\,\delta(q - q'). \qquad (16)$$

Substituting this expression into (8) and introducing the variables $Q = (q+q')/2$ and $\xi = q - q'$, we find

$$\frac{\partial \rho}{\partial t} = +\frac{i\hbar}{m}\frac{\partial^2 \rho}{\partial Q\,\partial \xi} - \frac{1}{i\hbar}\left\{V\left(Q + \frac{\xi}{2}\right) - V\left(Q - \frac{\xi}{2}\right)\right\}\rho. \qquad (17)$$

This form of the equation of motion for the density matrix

$$\rho(q, q') \equiv \rho(Q, \xi)$$

corresponds[1] in classical mechanics to the equations of motion for the Fourier transform of the classical density $\rho_c(Q, P)$ in phase space:

$$\rho_c(Q, \xi) = \int \rho_c(Q, P) \exp\left(-i\,\frac{\xi P}{\hbar}\right) dP \qquad (18)$$

(here $\hbar$ has so far been inserted purely formally). In fact, converting from $\rho_c(Q, P)$ to $\rho_c(Q, \xi)$ and putting

$$H(Q, \xi) = \int H(Q, P) \exp\left(-\frac{i\xi P}{\hbar}\right) dP, \qquad (19)$$

we get

$$i\hbar\,\frac{\partial \rho_c(Q, \xi)}{\partial t} = \int \left\{ uH(Q, u)\,\frac{\partial \rho_c(Q, \xi - u)}{\partial Q} - u\rho_c(Q, u) \right.$$
$$\left. \times \frac{\partial H(Q, \xi - u)}{\partial Q} \right\} du \qquad (20)$$

and in particular, when

$$H(Q, P) = \frac{P^2}{2m} + V(Q)$$

we get explicitly from (20) that

$$\frac{\partial \rho_c(Q, \xi)}{\partial t} = -\frac{i\hbar}{m}\,\frac{\partial^2 \rho_c(Q, \xi)}{\partial Q\,\partial \xi} + \frac{1}{i\hbar}\,\xi\,\frac{\partial V(Q)}{\partial Q}\,\rho_c(Q, \xi). \qquad (21)$$

We see on comparing this with the quantum equation (17) that the latter becomes (21) if

$$\left|\frac{\partial V}{\partial Q}\right| \gg \xi^2 \left|\frac{\partial^3 V}{\partial Q^3}\right|$$

i.e. if the potential $V(Q)$ is a sufficiently smooth function of the co-ordinates Q.

This shows that classical mechanics can also accommodate the method of describing the state of an ensemble via pairs of points q and q' in configuration space $R(q)$.

REFERENCE

1. D. Blokhintsev and Ch. Briskina, *Vestnik MGU* 10 (1942) 115.

CHAPTER VII

IS THE WAVE FUNCTION AVOIDABLE?

The basic equation of quantum mechanics – Schrödinger's equation – resembles the equations for the density matrix in that they can all be put in a form very similar to the equations of classical statistical mechanics or of the mechanics of a host of particles. This has given rise to a temptation that has more than once led to consideration[1] of quantum mechanics as a form of classical mechanics supplemented by some novel quantum force. The present author himself once hoped that the striking similarities between the equations for the $R(q, p)$ matrix and the equations of classical physics would allow the formulation of quantum mechanics as the statistical mechanics of quantities not simultaneously measurable. The basis of the theory would then be not the wave function but the $R(q, p)$ matrix, the analogue of the classical density $\rho_c(q, p)$ in $R(q, p)$ phase space. All such attempts employ, in one way or another, equations that are (in wave-function language) non-linear with respect to the wave function. This feature has given rise to a serious obstacle to such attempts.

Let us consider first what has been called the "hydrodynamic" form of Schrödinger's equation. Here the theory is based on the probability density $\rho(x)=|\psi(x)|^2$, with $\theta(x)=\arg \psi(x)$ as the phase of the wave function. Each of these has a classical interpretation: $\rho(x)$ may be considered as the mean density of the particles while $\nabla\theta(x)/m$ may be considered as the mean speed of these at the point x, so that $J=\rho\nabla\theta/m$ has the meaning of particle flux density. However, the equation for these quantities are non-linear; if in Schrödinger's equation

$$i\hbar \frac{\partial \psi}{\partial t} = -\frac{\hbar^2}{2m} \nabla^2\psi + V\psi \tag{1}$$

(for simplicity considered for one particle) we insert a wave function of the form

$$\psi = \rho^{\frac{1}{2}} \exp(i\theta/\hbar)$$

41

we obtain two equations for ρ and θ:

$$\frac{\partial \rho}{\partial t} + div\left(\rho \, \frac{1}{m} \, \nabla\theta\right) = 0 , \tag{2}$$

$$\frac{\partial \theta}{\partial t} + \frac{1}{2m}(\nabla\theta)^2 + V + \frac{\hbar^2}{m}\left(\frac{\nabla^2\rho}{\rho} + \frac{(\nabla\rho)^2}{\rho^2}\right) = 0 . \tag{2'}$$

Both may be considered as classical; the first as the equation of continuity for the particle density ρ and for the flux $J = \rho\nabla\theta/m$, and the second as the equation of the action function θ. The second equation coincides with the classical equation if the "quantum" term

$$\frac{\hbar^2}{m}\left(\frac{\nabla^2 p}{\rho} + \frac{(\nabla\rho)^2}{\rho^2}\right)$$

is considered as an additional "quantum" potential dependent on ρ and on derivatives of the latter.

At first sight this interpretation of the equations appears to be free from conflict, while the equations themselves are equivalent to the initial Schrödinger equation of (1). However, this is not so, because the principle of superposition of states plays a fundamental part in quantum mechanics. This principle indicates that, if there are two states of the quantum ensemble represented by wave functions $\psi_{\mathcal{M}_1}$ and $\psi_{\mathcal{M}_2}$, we can produce a macroscopic setting $\mathcal{M} = \mathcal{M}_1 + \mathcal{M}_2$ such that the new ensemble will be represented by the wave function

$$\psi_{\mathcal{M}} = C_1\psi_{\mathcal{M}_1} + C_2\psi_{\mathcal{M}_2} , \tag{3}$$

in which C_1 and C_2 are arbitrary numbers subject only to the normalization condition

$$|C_1|^2 + |C_2|^2 = 1 .$$

This simple but important principle cannot be expressed in the language of ρ and θ without resort to the wave function again. An even greater inconvenience arises in formulating the symmetry conditions for a system of identical particles. These conditions for the wave function state that

$$\mathcal{P}_{ik}\psi = \pm\psi , \tag{4}$$

in which $\mathscr{P}_{ik}$ is the interchange operator for particles i and k, while the sign $\pm$ depends on the statistics (Fermi or Bose-Einstein). This simple condition is formulated in a clear fashion in the linear apparatus of quantum mechanics, but it is reflected only in a very obscure fashion in the language of ρ and θ. For instance, suppose we have two identical particles; then we may consider (1) and (2) as equations of functions describing the relative motion of these particles, while the motion of the centre of gravity is distinguished in the usual way by separating the variables. Symmetry condition (4) is then transferred to the function for the relative motion (the function describing the motion of the centre of gravity is automatically symmetrical with respect to exchange); it means that $\psi(x)$ must be even or odd when x is replaced by $-x$, and (4) thus implies that

$$\rho(-x) = \rho(x), \quad \theta(-x) = \theta(x) + \pi S, \tag{5}$$

in which $S = 0, 2, 4, \ldots$ or $S = 1, 3, 5, \ldots$.

The condition on $\rho(x)$ is mathematically of no inconvenience, but the second condition is extremely troublesome; if we seek ways of meeting this requirement, we find that the simplest way of satisfying (5) is to return to the wave function and the condition that this be even or odd.

Another form for the quantum-mechanical equations is close to that of classical statistical mechanics and is based on the use of the $R(q, p)$ density matrix. Formulae (V–15), (V–16), and (VI–12) show that there is a close analogy between the equations of classical statistical physics and the equations for the quantum density $R(q, p)$. However, we have seen (Chapter V) that matrix $R(q, p)$ differs from the classical density in being obliged to have a non-zero imaginary part, so in the simple sense of the word there cannot be a positively definitive probability or density in $R(q, p)$ phase space. The complex character of $R(q, p)$ is an expression of the principle of complementarity, but it might have been hoped that $R(q, p)$ could be considered as an extension of the probability density concept to the case of variables q and p not simultaneously measurable, this matrix being used as the basis of the theory instead of the wave function. However, the matrix $R(q, p)$ is non-linear with respect to the wave function, so all the difficulties noted above for the $\rho - \theta$ description also occur here. In particular, the difficulty over (3) (principle of super-position) that arose with ρ and θ persists when R is used.

In addition, it is extremely difficult to meet the hermiticity conditions (V–19) and the symmetry conditions (V–20) for the case of identical particles unless we resort to the wave-function concept; it is probable that they may be satisfied in general form only by considering $R(q, p)$ as bilinear in $\psi^*(p)$ and $\psi(q)$.

These difficulties are common to all attempts to formulate quantum mechanics in the language of quantities non-linear with respect to the wave function; throw the latter out of the door and it will come back through the window. This persistence of the wave function is an expression of the fact that the language of the linear theory is that of the essential nature of quantum mechanics; translation into any other tongue will always cause the poetry of this linearity to sound weaker than in the original.

REFERENCE

1. *Causality in Quantum Mechanics* [Russian translation], IL, 1955.

CHAPTER VIII

IS THE WAVE FUNCTION MEASURABLE?

We have defended the use of the wave function, but we now must consider what it represents and whether it can be measured or established by experiment, as for example a particle density or the temperature of a gas. Frequently the answer is given in the negative, as the wave function is essentially undefined as regards a phase constant; the two wave functions ψ and ψ' related by

$$\psi' = \psi e^{i\alpha}, \tag{1}$$

(in which α is a real number) represent the same quantum ensemble. Hence in speaking of the wave function as measurable or otherwise, we will from the start not demand the impossible (or rather meaningless) and will understand by measurement of the wave function merely measurement apart from the phase constant α.

We may avoid a complicated argument by considering the function in explicit form in terms of the coordinates x of one particle (by x we might also understand the relative coordinates of two particles). The wave function is complex in the general case, and we put it in the form

$$\psi(x) = |\psi(x)|\, e^{i\theta(x)}, \tag{2}$$

in which $|\psi(x)|$ is the real amplitude and $\theta(x)$ is the phase. There is no fundamental objection to the experimental determination of $\rho(x)$, the mean particle density, so this amplitude may be determined, since by definition

$$\rho(x) = |\psi(x)|^2, \quad |\psi(x)| = +\,\rho(x)^{\frac{1}{2}}, \tag{3}$$

Further, the mean current density may also be measured:

$$\mathbf{J} = \frac{\hbar}{m}\,\rho\nabla\theta. \tag{4}$$

Then $\theta(x)$ is given by

$$\theta(x) = \frac{m}{\hbar}\int_{x_0}^{x}\frac{\mathbf{J}}{\rho}\,\mathrm{ds}, \tag{5}$$

in which **ds** is a length element and x_0 is some arbitrary point, the arbitrariness arising from the arbitrary choice of the phase in (1). For clarity we suppose that we are considering the measurement of $\rho(x)$ and $\theta(x)$ in an atom or ion with one electron in an s-state or a p-state. In the first case the wave function has spherical symmetry and currents are absent, so

$$\psi(x) = \sqrt{\rho(x)} \tag{6}$$

and measurement of $\rho(x)$ gives us directly the values of the wave function.

In the second case the wave function assumes the following form in polar coordinates r, ϑ, ϕ:

$$\psi(r, \vartheta, \phi) = A(r, \theta) e^{im\phi}, \quad m = 0, \pm 1, \tag{7}$$

the double sign corresponding to the two orientations of the atom in a magnetic field. We assume that this field is used to produce one of the possible orientations; then we measure $\rho(x)$ to find the modulus of the function:

$$A(r, \theta) = \sqrt{\rho(r, \theta)}, \tag{8}$$

while by measuring the magnetic moment of the atom, which is

$$\mathscr{M} = \frac{e\hbar}{2\mu c} m, \tag{9}$$

(μ being the mass of an electron) we find also the phase $\theta = m\phi$ of the wave function. These simple examples show that the wave function can be measured.

The above examples concern measurements on bound states. Consider now the case of particle scattering, where the wave function takes the form

$$\psi(x) = e^{ikx} + u(x), \tag{10}$$

in which $\exp(ikx)$ is the primary incident wave (whose wave vector is $k = 1/\lambda$) and $u(x)$ is the scattered wave, the latter taking the following form at great distances from the scattering point:

$$u(x) = \frac{e^{ikr}}{r} A(k, \vartheta), \tag{11}$$

$$A(k, \vartheta) = \frac{1}{2ik} \sum_{0}^{\infty} (e^{2i\eta_l} - 1) \mathscr{P}_l(\cos\vartheta), \tag{12}$$

in which $\eta_l(k)$ are the phases of the scattered waves and $\mathscr{P}_l(\cos \vartheta)$ are Legendre polynomials. We assume that we measure the particle density (although it would be nearer to the practical case to speak of measuring the particle flux density), which is

$$|\psi|^2 = |\psi_0 + u|^2$$

and which contains the interference term

$$\psi_0 u^* + \psi_0^* u,$$

which complicates the argument. However, the difficulty is readily avoided by realizing that description of the incident wave as a plane wave represents an abstraction, since strictly speaking a beam bounded in space (a wave packet) should be considered. Then we observe particles scattered far away from the primary beam, thus avoiding interference between the primary and scattered waves (note that these waves are coherent).

This is a realistic representation of an experiment for measuring $|u|^2$ in the usual way; experiment thus gives us

$$|u|^2 = \frac{1}{r^2} |A(k, \vartheta)|^2. \tag{13}$$

Hence we find the amplitude of the scattered wave $A(k, \vartheta)$, but not the phase; in general,

$$A(k, \vartheta) = |A(k, \vartheta)| e^{i\alpha(k, \vartheta)}$$

and the phase $\alpha(k, \vartheta)$ is lost from the measured quantity; an additional measurement of the flux density of the scattered particles yields nothing new, since the formula for the flux density

$$J = \frac{i\hbar}{m} (u^* \nabla u - u \nabla u^*) \tag{14}$$

implies that

$$J = \frac{\hbar k}{m} |u|^2$$

so J is simply proportional to $|u|^2$. The phase uncertainty gives rise to the serious and troublesome problem of recovering the phase from the experimental data, by which is meant the part of the phase $\alpha(k, \vartheta)$ that is dependent on the angle ϑ and momentum k. The problem may be illustrated by reference to two particular cases. First we assume that only the first phase differs from zero (S-scattering); in this very simple case we get from (12) that

$$|A(k,\eta)|^2 = \frac{1}{4k^2} \sin^2 \eta_0 \,, \tag{15}$$

$$\sin \eta_0 = \pm \, |A(k,\eta)| \, 2k \,, \tag{15'}$$

and so the phase of the S-scattering is determined apart from sign. Here there thus arises an ambiguity, which is not entirely irrelevant for it conceals, for example, the question whether the forces between the particles are ones of repulsion or attraction.

This phase η_0 is related[1] to the particle interaction potential $V(r)$ (for weak interaction) as follows:

$$\eta_0 = -\frac{2mk}{\hbar^2} \int_0^\infty V(r) \left[\left(\frac{\pi}{2kr} \right)^{\frac{1}{2}} I_{\frac{1}{2}}(kr) \right]^2 r^2 \, dr$$

in which $J_{\frac{1}{2}}(z)$ is a Bessel function, so that the sign of $V(r)$ determines that of η_0.

Consider now the more complicated case where both phases are different from zero, e.g. the S-phase and the $\mathscr{P}$-phase. Here

$$|A(k,\vartheta)|^2 = \frac{1}{4k^2} \left\{ \sin^2 \eta_0 + [2\cos(\eta_0 - \eta_1) - 2\cos 2\eta_0 - 2\cos 2\eta_1] \cos \vartheta \right.$$

$$\left. + \sin^2 \eta_1 \cos^2 \vartheta \right\}. \tag{16}$$

The coefficients of $\cos^0 \vartheta$, and $\cos^1 \vartheta$, and $\cos^2 \vartheta$ may be determined to give us the phases η_0 and η_1 apart from sign.

These simple examples show that the problem of recovering the phase of the wave function (even the asymptotic value) is far from simple, although the difficulties are not fundamental. This is illustrated by the observation that for (say) proton-proton scattering we may deduce the sign of η_0 from the interference between the nuclear scattering and the scattering arising from the Coulomb field of the nucleus. The differential cross-section for elastic scattering of protons into a solid angle $d\Omega$ (after averaging with respect to the spins, see ref. 2) is as follows when the specifically nuclear interaction is included:

$$d\sigma(\vartheta) = \frac{e^4}{E^2}\cos\vartheta\left[\frac{1}{\sin^4\vartheta} + \frac{1}{\cos^4\vartheta} - \frac{1}{\sin^2\vartheta}\cdot\frac{1}{\cos^2\vartheta} - \frac{\hbar v}{e^2}\frac{\sin^2\eta_0}{\sin^2\vartheta\cos^2\vartheta} + \frac{2\hbar v}{e^2}\sin^2\eta_0\right]d\Omega,\tag{17}$$

in which E is proton energy, v is the relative velocity of the protons, e is the electronic charge, and η_0 is the phase of the nuclear interaction. We may assume that this cross-section is known from experiment, and consequently that we know the correction to the pure Coulomb scattering arising from the nuclear interaction in the S-state. This correction is

$$f(E,\vartheta)\,d\Omega = -\frac{e^4}{E}\cos\vartheta\left[\frac{\hbar v}{e^2}\frac{\sin^2\eta_0}{\sin^2\vartheta\cos^2\vartheta} - \left(\frac{2\hbar v}{e^2}\sin^2\eta_0\right)\right]d\Omega.\tag{18}$$

We differentiate $f(E,\vartheta)$ with respect to ϑ for fixed energy to get

$$\frac{\partial f(E,\vartheta)}{\partial\vartheta} = -\frac{e^4}{E^2}\frac{\hbar v}{e^2}\sin^2\eta_0\frac{d}{dv}\left[\frac{1}{\sin^2\vartheta\cos^2\vartheta}\right],\tag{18'}$$

which defines η_0. Allowance must be made for higher phases at higher proton energies, which makes the problem more complicated.

These simple examples show that the wave function can be measured.

Let us now consider in more detail the measurements actually needed, first in relation to particle scattering, which for simplicity we assume to be elastic. The directly measured quantity is the differential cross-section which is related to the scattering amplitude by

$$\sigma(\vartheta,k) = |A(\vartheta,k)|^2.\tag{19}$$

We have just discussed the determination of the amplitude itself, so we shall now concentrate attention on the determination of the cross-section σ. From its definition we have

$$dN = \sigma\cdot J\,\frac{dS}{r^2} = \sigma\cdot J\,d\Omega,\tag{20}$$

in which J is the current density in the incident wave and dN is the number of particles scattered in unit time into an element of solid angle

$d\Omega$. This implies directly that it is insufficient to observe a single act of scattering in order to determine σ, for this tells us exactly nothing about $\sigma(\vartheta, k)$ as a function of the scattering angle ϑ. This single act of scattering may be typical (probable) or atypical (improbable). We must observe many acts of scattering at various angles in order to obtain $\sigma(\vartheta, k)$ as a function of ϑ from experiment, and the numbers must be such as to rule out errors arising from random fluctuations. This is essentially a trivial feature known to any experimentalist, and it needs to be mentioned only because it is stated in many textbooks on quantum mechanics that the wave function is a characteristic of the state of a single particle. If this were so, it would be of interest to perform such a measurement on a single particle (say an electron) which would allow us to determine its own individual wave function. No such measurement is possible.

This contradiction is abolished if we bear in mind the idea that the wave function is a characteristic arising from the existence of a quantum ensemble for the particles, i.e. if we consider a particle μ together with its macroscopic setting $\mathscr{M}$, the latter governing the conditions of motion of the former.

Now we may turn to measurement of the wave function for a bound state; here we encounter some features that deserve attention.

In the above experiment on the asymptotic behaviour of the wave function in elastic scattering, the number of scattered particles was measured directly by some macroscopic device, e.g. a Geiger counter (Figure 6), which was entirely feasible in that case because it was a

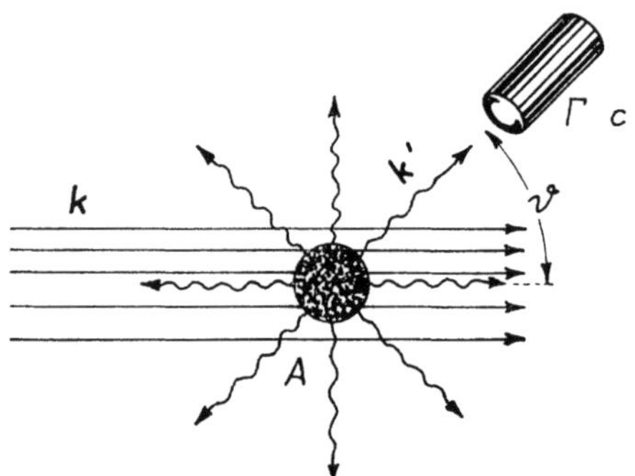

Fig. 6. The primary wave **k** is scattered by a microsystem A, the scattered wave **k′** being shown by wavy lines. C is a counter for particles scattered at an angle ϑ.

question of observing free scattered particles far from the scattering centre, i.e. far from the site of all the microscopic processes. This type of measurement can be called direct. The situation is fundamentally different for a bound state: here the wave function is localized in a microscopic volume, within which no macroscopic measuring device can be inserted. Hence any measurement of the wave function in this case must be based on some means of indirect measurement[3] via a directly measurable interaction of the system with some other system. The last may be a free particle scattered by the system.

Hence the indirect measurement of the wave function for a bound state (for which no direct measurement exists) reduces to our previous experiment on the scattering of free particles. It should hardly be necessary here to stress again that no single measurement can give us any information on the wave function of the bound state. We must find the cross-section of the scattered particle as the first step in our indirect measurement, since this is used as our probe to study the innards of the atomic system. Here we have to ensure (in experimentalists' terminology) good statistics, i.e., we have to make many measurements on the scattering of our probe particle by the system.

We deduce the differential cross-section for the scattering from direct measurements of the number of scattered probe particles; the wave function of the system must be deduced via recovery of the scattering amplitude $A(\vartheta, k)$.

The above arguments have shown that this problem often does not allow an unambiguous solution, but if we assume that we have been able to determine the scattering amplitude for the probe particle then all the information on the atomic system is contained in the infinite set of phases $\eta_0(k)$, $\eta_1(k), \ldots, \eta_e(k)$, which are functions of the energy of the relative motion of the probe particle and the atomic system. Even in this very optimistic case we encounter the complicated mathematical task of deriving the structure of the object from its optical image, which is now considered as specified via the amplitude $A(\vartheta, k)$. Of course, this is provided by experiment as the angular intensity distribution for the scattered particles, or (which is the same) as the differential cross-section; it should be stressed that this mathematical problem, no matter how complicated, is soluble, at least in certain cases, which of itself shows that in principle we can pursue the path of indirect measurement.

A bound state is a state of an isolated microsystem, so it will be more convenient in what follows to speak of studying the structure of an atomic microsystem (we have already used this terminology), since in the general case we examine not only the wave function of the system but also the mode of interaction with the probe particle. The next chapter deals with reconstruction of structure from the image of the microsystem.

REFERENCES

1. N. Mott and H. Massey, *The Theory of Atomic Collisions*, 2nd ed., 1949.
2. A. Akhiezer and I. Pomeranchuk, *Some Aspects of Nuclear Theory* [in Russian], GITTL, 1948.
3. L. I. Mandel'shtam, *Lectures on Quantum Mechanics*. Collected Works [in Russian], vol. 5, Izd. AN SSSR, 1950.

CHAPTER IX

DEDUCTION OF THE STRUCTURE OF A MICRO-OBJECT FROM PARTICLE SCATTERING

The basis of our discussion of this topic is Schrödinger's equation for the probe particle. Let m be the reduced mass of this particle, which is found from $1/m = 1/\mu + 1/\mathcal{M}$, in which μ is the actual mass of the scattered particle and $\mathcal{M}$ is the mass of the scatterer, and let x be the coordinate of the scattered particle relative to the centre of gravity. Then Schrödinger's equation for the relative motion $\psi(x)$ is as follows, any motion of the centre of gravity being of no interest:

$$\frac{\hbar^2}{2m}\nabla^2\psi(x) + V(x)\psi(x) = E\psi(x), \tag{1}$$

in which $V(x)$ is a function describing the interaction between the micro-system and the probe particle whose energy is E. The wave function must take the form

$$\psi(x) = e^{ikz} + u(x), \tag{2}$$

in which $\exp(ikz)$ is the primary plane wave and $u(x)$ is the scattered wave, the latter for large distances $z = |x| \to \infty$ being of the form

$$u(x)_{r\to\infty} = \frac{e^{ikr}}{r} A(k, \vartheta), \tag{3}$$

where $A(k, \vartheta)$ is the amplitude, which from (VIII–19) determines the differential scattering cross-section, i.e. the experimentally observed quantity.

By determination or deduction of the structure we shall understand determination of $V(x)$ from data on the scattering of the probe particle. The term "structure of the object" is of optical origin. We put (2) in the form

$$\nabla^2\psi(x) + k^2 n^2(x)\psi(x) = 0, \quad n^2(x) = \left[\frac{E - V(x)}{E}\right], \tag{4}$$

and see that $n^2(x)$ may be considered as the refractive index of the scattering medium formed by our microsystem.

Consider the case of weak scattering, when the primary wave is only slightly altered within the scattering region. In this particularly simple case we may apply the Born approximation, the essential feature of which is that $\psi(x)$ in $V(x)\,\psi(x)$ is replaced by the primary wave undistorted by scattering, $\psi_0(x) = \exp(ikz)$. This allows (2) to be written in the form of an equation with a given right-hand side:

$$\nabla^2 u + k^2 u = \frac{2m}{\hbar^2}\,V(x)\,\psi_0(x). \tag{5}$$

The equation for the Green's function of this equation is

$$\nabla^2 G + k^2 G = -\,\delta(x - x'), \tag{6}$$

which has a solution for divergent waves of the form

$$G = -\frac{1}{4\pi}\,\frac{e^{ik|x-x'|}}{|x - x'|}. \tag{7}$$

Hence the solution to the inhomogeneous equation (5) is

$$u(x) = -\frac{1}{4\pi}\,\frac{2m}{\hbar^2}\int V(x')\,\psi_0(x')\frac{e^{ik|x-x'|}}{|x - x'|}\,\mathrm{d}^3x', \tag{8}$$

which for $|x-x'| \to \infty$ becomes

$$u(x) = -\frac{e^{ikr}}{2\pi r}\,\frac{2m}{\hbar^2}\int V(\mathbf{x}')\,e^{i\mathbf{q}\mathbf{x}'}\,\mathrm{d}^3x', \tag{9}$$

where $\mathbf{q} = \mathbf{k}' - \mathbf{k}$ is the change in the wave vector due to the scattering; $|q| = 2k\sin(\vartheta/2)$, in which ϑ is the angle between the direction of the primary wave and that of the scattered wave (Fig. 6). Comparison with (3) shows that the amplitude of the scattered wave is

$$A(k, \vartheta) = A(q) = -\frac{1}{4\pi}\,\frac{2m}{\hbar^2}\int V(x)\,e^{i\mathbf{q}x}\,\mathrm{d}^3x. \tag{10}$$

This shows that the Born approximation gives the amplitude of the scattered wave simply as the Fourier component of $V(x)$.

Let us now consider the deduction of this last function from data on the scattering amplitude $A(q)$, which we put in the form

$$A(q) = a(q)\,e^{i\alpha(q)}$$

in which

$$a(q) = |A(q)|$$

and $\alpha(q)$ is the phase; we also note that $V(x)$ is taken as being real (no absorption, only elastic scattering). We then have formally from the inverse Fourier transform that

$$V(x) = -\frac{8\pi m}{h^2} \int A(q) e^{i\mathbf{q}\mathbf{x}} \mathrm{d}^3 q, \tag{11}$$

in which the integral with respect to the transferred momentum is taken between infinite limits.

The first restriction on $V(x)$ arises because we know the amplitude only for $q < 2k$ from measurements at probe-particle energies not exceeding E, with $k^2 = 2mE/h^2$, so that we are obliged to replace (11) by another integral within the limits $|q| < 2k$. Consequently we obtain in place of the true $V(x)$ an imprecisely defined representation $\bar{V}(x)$:

$$\bar{V}(x) = \int\limits_{q < 2k} A(q) e^{i\mathbf{q}\mathbf{x}} \mathrm{d}^3 q. \tag{12}$$

This expresses the fact that we cannot see details of any object whose size a is smaller than the wavelength λ of the "illumination".

The second restriction arises because the phase $\alpha(q)$ cannot be determined; the transform of (12) shows that this phase happens to be zero for the case where the interaction function is symmetric with respect to the inversion $\mathbf{x} \to -\mathbf{x}$. However, the interaction may contain a part not symmetric with respect to this transform, in which case $\alpha(q)$ is not zero, so it is not possible to determine $V(x)$ uniquely from the differential scattering cross-section alone. The utility of (12) is therefore very much dependent on the assumptions about the symmetry of the interaction that may be made *before the measurement*.

Consider now the relation of structure deduction to the measurement of the wave function for a bound state. We assume that the investigated micro-object is a heavy ion, or an atom with one electron. The wave function of this electron within the atom is denoted by $\varphi_0(x)$. The mean charge density this produces at a point x (in which x is the coordinate of the electron relative to the nucleus) is

$$e\rho(x) = e|\varphi_0(x)|^2 ;$$

If the nucleus has a charge $+ eZ$, the potential energy of the second electron (probe particle) in the field produced by the nucleus and the first electron is

$$V(x) = -\frac{e^2 Z}{r} + e^2 \int \frac{|\varphi_0(x')|^2}{|x - x'|} \, d^3 x'. \tag{13}$$

The Fourier transform of this is

$$\tilde{V}(q) = -\frac{4\pi e^2 Z}{q^2} + \frac{4\pi e^2}{q^2} \, \tilde{\rho}(q), \tag{14}$$

in which $\tilde{\rho}(q)$ is the Fourier transform of $\rho(x)$.

If $\rho(x)$ is even with respect to the inversion $x \to -x$, we get from the above argument that we may determine an approximate representation of $\bar{\rho}(x)$:

$$\bar{\rho}(x) = \int\limits_{q < 2k} \rho(q) e^{iqx} \, d^3 q. \tag{15}$$

If it is also known from considerations of symmetry that $\varphi_0(x)$ is real, we get the approximate transform of this function as

$$\bar{\varphi}_0(x) = \sqrt{\bar{\rho}(x)}. \tag{16}$$

Thus we see that the wave function is in principle measurable for a bound state, at least in certain very simple cases when the indirect measurements may be interpreted unambiguously. In every case it is first necessary to measure the differential cross-section for elastic scattering of probe particles, i.e. we must provide good statistics for the acts of scattering, which implies repeated scattering of an electron at the atom in the specified initial state. At first sight it might seem that in this series of repeated scattering experiments we could use, at least in principle, a single atom as the target for our probe particles, but this is not so, for the essence of the matter is that we must use a probe beam of wavelength $\lambda = 1/k \ll a$ to study the structure of a target whose dimensions are of the order of a. The binding energy of a particle in this object is $I \approx \hbar^2/2ma^2$, i.e. the kinetic energy of the probe particle is $E = \hbar^2 k^2/2m \gg I$, so that the atom will be ionized by the flux of probe particles and hence must be replaced.

This argument needs some detailed revision when the probe particle mass m differs from the mass μ of the intra-atomic particle; we still have

56

$E \gg I$ if $m < \mu$, but for $m > \mu$ we must remember that the Born approximation becomes unsuitable for $v < v_0$, in which v is the speed of the probe particle and v_0 is the velocity of the atomic particle, e.g. of an electron in an atom. We have $v_0 \approx \hbar/a\mu$ roughly, so $E > I$ is replaced by $E > mI/\mu$. Thus E exceeds the ionization energy of the atom, which is therefore ionized by irradiation. In the language of photography, we might say that the atom is too restless a subject to give a good portrait – its face changes in the most inexplicable fashion during the exposure.

The situation may be described quantitatively as follows. Let σ_0 be the cross-section for elastic collisions and σ_i that for inelastic ones; then the ratio of the number of acts of inelastic scattering N_i to the number N_0 of elastic ones will be Q_i/Q_0. A good "photograph" requires that $N_i \ll N_0$, but this condition cannot be met because the condition $E \gg I$ implies that the electron in the atom may be considered as almost free in relation to the probe particle, since its binding energy I is small relative to the kinetic energy of the incident probe particle. Then σ_i exceeds σ_0; for $E \gg I$ we have[1]

$$\sigma_i = 2\pi a^2 I \log \frac{E^3}{2I^2 m}, \quad \sigma_0 = \frac{7\pi}{3} a^2 \frac{I}{E}, \tag{17}$$

so that $N_0/N_i = \sigma_0/\sigma_i \to 0$ for $E \gg I$.

This makes it impossible to obtain a sharp image of the atom by irradiating a single atom. Hence it is clearly impossible to deduce the wave function of the electron in the atom if we probe the same atom, when, as we have seen, there is no guarantee that the atom will not be ionized during the measurement.

This does not mean that it is essentially impossible to observe a single atom on, say, the stage of an electron microscope. If we restrict ourselves to this less ambitious problem of an atom on some stage and neglect any possible changes in the atom during inelastic processes, we may (at least for fairly heavy atoms) perform several thousand acts of scattering[2] without displacing the atom from its initial position of attachment (*adsorption*).

REFERENCES

1. N. F. Mott and H. S. W. Massey, *The Theory of Atomic Collisions*, 2nd ed., 1949.
2. D. Blokhintsev, *ZhETF* **17** (1947) 814.

CHAPTER X

THE INVERSE PROBLEM IN QUANTUM MECHANICS

The usual problem in quantum mechanics is that of finding the proper wave functions $\psi(x, E)$ and proper values E of the Hamilton operator $\mathcal{H}(x)$ (assumed to be known); this is the direct problem. However, we may also conceive the inverse problem, in which we have experimental data on the energy states E and proper functions $\psi(x, E)$ but wish to find $\mathcal{H}(x)$. The mode of solution has already been presented for the case where the Born approximation may be used. Several recent mathematical works [1,2,3,4] have presented a method of solving this problem without demanding the assumption of weak interaction required by the Born approximation (Chapter IX).

The method may be illustrated by reference to scattering in the S state for a system whose hamiltonian is

$$\mathcal{H}(x) = -\frac{\hbar^2}{2m}\nabla^2 + U(r).\qquad(1)$$

Denoting the wave number by

$$k = \sqrt{\frac{2mE}{\hbar^2}}$$

and putting

$$\frac{2m}{\hbar^2}\,U(r) = V(r)$$

we obtain the Schrödinger equation for the S wave $\varphi(r, k)$, as

$$-\frac{d^2\varphi}{dr^2} + V(r)\varphi = k^2\varphi\qquad(2)$$

subject to the boundary conditions

$$\varphi(0, k) = 0, \quad \frac{d\varphi(0, k)}{dr} = k\qquad(3)$$

and to the assumption that

$$\int_0^\infty r|V(r)|dr < \mathcal{M}$$

(convergence of the integral). The form of $\varphi(r, k)$ for $r \to \infty$ is

$$\varphi(r, k) = A(k) \sin(kr + \delta(k)), \tag{4}$$

in which $A(k)$ is the amplitude of the scattered wave and $\delta(k)$ is the phase; the latter is observable, because the total cross-section σ for the elastic S scattering is

$$\sigma(k) = \frac{\pi}{k^2} \sin^2 \delta(k). \tag{5}$$

As regards $A(k)$, we assume for simplicity that the system has no bound states (the following relation becomes rather more complicated if these are present), and this amplitude may be found from the dispersion relation

$$\log A(k) = \frac{1}{\pi} \int_{-\infty}^{+\infty} \frac{\delta(k')\,dk'}{(k' - k)}. \tag{6}$$

This relation is readily derived if we consider a solution whose behaviour as $r \to \infty$ is of the following form instead of the $\varphi(r, k)$ with the asymptotic behaviour as of (4):

$$u(r, k) = \frac{A(k)}{2i} e^{i[kr + \delta(k)]} \tag{4'}$$

the boundary condition being

$$u(0, k) = \frac{1}{2i}, \quad u'(0, k) = \frac{k}{2}. \tag{3'}$$

We then have that

$$f(k) = \frac{2i}{A(k)} e^{-i\delta(k)} u(0, k),$$

$$S(k) = e^{2i\delta(k)} = \frac{f^*(k)}{f(k)}, \tag{7}$$

in which $S(k)$ is the scattering matrix for the S wave; (7) implies that

$$\operatorname{Im} \log f(k) = 2\delta(k), \tag{8}$$

and since $f(k)$ is an analytical function (in the absence of bound states) for $\operatorname{Im} k > 0$ and

$$\operatorname{Re} \log f(k) = -A(k),$$

so that from (8) we apply Cauchy's theorem to the boundary consisting of an infinitely large semicircle and the real axis, evading the point k, and hence arrive at (6).

The solution to (2) that satisfies the conditions of (3) may be put as

$$\varphi(r, k) = \sin kr + \int_0^r K(r, t) \sin kt \cdot dt, \tag{9}$$

in which the kernel $K(r, t)$ for $0 \leqslant t \leqslant r$ satisfies

$$\frac{\partial^2 K(r, t)}{\partial r^2} - \frac{\partial^2 K(r, t)}{\partial t^2} = V(r) K(r, t) \tag{10}$$

subject to the conditions

$$K(r, 0) = 0, \quad K(r, r) = \tfrac{1}{2} \int_0^r V(z)\,dz, \tag{11}$$

as may be verified by direct substitution of (9) into (2). Here (11) allows us to eliminate the potential $V(r)$ from (10):

$$V(r) = 2\,\frac{dK(r, r)}{dr}. \tag{12}$$

However, we will not do this at present but will set out the solution to (10), which is readily obtained by successive approximation:

$$K(r, t) = \tfrac{1}{2} \int_{(r-t)/2}^{(r+t)/2} V(z)\,dz + \tfrac{1}{2} \sum_{n=1}^{\infty} \int_{(r-t)/2}^{(r+t)/2} du_n \int_0^{(r-t)/2} dv_n\, V(u_n + v_n)$$

$$\int_0^{u_n} du_{n-1} \int_0^{v_n} dv_{n-1}\, V(u_{n-1} + v_{n-1}) \ldots \int_{u_1}^{v_1} V(z)\,dz. \tag{13}$$

60

This gives us the partial derivative with respect to r as

$$\frac{\partial K(r,t)}{\partial r} = -\frac{\partial K(r,t)}{\partial t} + \tfrac{1}{2} V\left(\frac{r+t}{2}\right)$$

$$+ \tfrac{1}{2} \sum_{n=1}^{\infty} \int_0^{(r-t)/2} dv_n V\left(\frac{r+t}{2} + v_n\right) \int_0^{(r+t)/2} du_{n-1} \qquad (14)$$

$$\cdot \int_0^{v_n} dv_{n-1} V(u_{n-1} + v_{n-1}) \ldots \int_{v_1}^{u_1} V(z)\,dz .$$

Further we formulate $\partial\varphi(r,k)/\partial r$ from (9) and replace $\partial K(r,t)/\partial r$ by the series of (14); integrating by parts, we get

$$\frac{\partial\varphi(r,k)}{\partial r} = k\left(\cos kr + \int_0^{-r} K(r,t)\cos kt\,dt \right)$$

$$+ \tfrac{1}{2} \int_0^{r} V\left(\frac{r+t}{2}\right)\sin kt + \tfrac{1}{2} \sum_{n=1}^{\infty} \sin kt \left\{ \int_0^{(r+t)/2} dv_n V\left(\frac{r+t}{2} + v_n\right)\right\}$$

$$\int_0^{(r+t)/2} du_{n-1} \int_0^{(r-t)/2} dv_{n-1} V(u_{n-1} + v_{n-1}) \ldots \int V(z)\,dz . \qquad (15)$$

Next we consider the expression

$$\left[\frac{1}{k}\frac{\partial\varphi(r,k)}{\partial r} + i\varphi(r,k) \right] e^{-ikr}$$

for $r\to\infty$. It can be shown quite simply from (9) and (15) that this limit is

$$1 + \int_0^{\infty} K_0(u)\, e^{-iku}\, du ,$$

in which $K_0(u)$ is the limit of $K(r,t)$ as r and t tend to infinity along the straight line $r-t=u>0$. On the other hand, the initial conditions of (3) and the asymptotic expression of (4) for $\varphi(r,k)$ together imply that this limit is

$$A(k) \exp[i\delta(k)],$$

so that

$$1 + \int_0^\infty K_0(u)\, e^{-iku}\, \mathrm{d}u = A(k)\, e^{i\delta(k)} = \psi(k). \tag{16}$$

$K(r, t)$ may be produced with respect to t in the odd way for $t < 0$ ($-r \leqslant t \leqslant 0$), while $\psi(k)$ may be produced for $k < 0$ in a way such that

$$A(k) = A(-k) \quad \text{and} \quad \delta(k) = -\delta(-k).$$

It can be shown[2] that (10) and (11) together have a unique solution within the triangle $-r \leqslant t \leqslant 0$, $r \geqslant 0$ for which

$$K(r, t) = -K(r, -t) \quad \text{and} \quad K(r, t) \to K_0(u)$$

when r and t tend to infinity along the straight line $u = r - t$.

This shows that it is possible to find $K(r, t)$ if $K_0(u)$ is known, and further to deduce $V(r)$ from (11).

Unfortunately, experience has revealed little practical value in this mathematically very striking method of structure deduction from particle scattering in the S state (or in any other state with a definite momentum l) by reference to the phase $\delta_l(k)$. Essentially, experiment gives us $\delta(k)$ only for a finite range $0 < k < k_{\max}$, not for the full range $0 < k < \infty$, whereas the behaviour of $V(r)$ is very sensitive to the behaviour of $\delta(k)$ at large k. This feature reflects a major law of optics: details in the object of scale a cannot be seen if the wavelength $\lambdabar$ of the "light" is greater than a.

We must have $\lambdabar$ less than a, i.e. $k \gg 1/a$, in order to examine the form of $V(r)$ for $r \approx a$, but $\delta(k)$ for elastic scattering becomes small for large k, and the simpler Born method may be applied.

In practice, attempts to examine nucleon interaction potentials encounter the difficulty that mesons start to be produced and relativistic effects become prominent for large k. The radius of action of nuclear forces is comparable with the Compton length $a = \hbar/mc$ of the meson (m is the meson's mass and c is the velocity of light). Hence a study of $V(r)$ as a nucleon interaction potential requires

$$k = \frac{1}{\lambdabar} \gg \frac{1}{a} = \frac{mc}{\hbar},$$

i.e. that the relative momentum of the nucleons is

$$p = \hbar k > mc.$$

This implies that the nucleon energy is

$$2E = \frac{2p^2}{M} \gg \frac{2m}{M} mc^2 \cong \tfrac{2}{7}mc^2,$$

$$v \gg \frac{mc}{M} = \tfrac{1}{7}c.$$

and their velocity is $v \gg mc/m = c/f$. This represents the threshold for meson production and also the range where relativistic effects become prominent, and neither of these effects has been incorporated in the non-relativistic theory presented here.

REFERENCES

1. I. M. Gel'fand and B. M. Levitan, *Izv. AN SSSR, ser. matem.* **15** (1951) 309.
2. V. K. Mel'nikov, *Uspekhi matem. nauk* **14** (1959) 121.
3. M. G. Krein, *Dokl. Akademii Nauk SSSR*, **105**, 3 (1955).
4. R. Jost and W. Kohn, *Phys. Rev.* **87**, no. 6 (1952).

CHAPTER XI

A MEASURING INSTRUMENT IS A MACROSCOPIC DEVICE

Quantum mechanics resembles classical mechanics in that the micro-systems are assigned various characteristic dynamic variables. Moreover, the situation may be formulated very crudely by saying that these variables have analogues in classical mechanics, such as particle co-ordinates, momentum, and total energy. There are exceptions in the case of certain specifically quantum variables, which have no analogues at all in classical theory, or else ones of very little significance, which include the particle spin σ and the isotopic spin τ. Most dynamic variables retain their significance in both forms of mechanics, at least in the sense that they become identical when the phenomena occur in a region where the two theories overlap. Formally such a region is characterized by the fact that any observable quantity L may be represented as a power series in Planck's constant $\hbar$:

$$L = L_0(p,q) + \sum_{s=1}^{\infty} \hbar^s \cdot a_s(p,q), \tag{1}$$

in which the first term corresponds to the classical value of L.

Some phenomena make this expansion impossible, and here the dynamic variables split up into complementary sets: space-time ones (q) and momentum-energy ones (p). This split into two complementary classes is a fundamental feature of quantum mechanics and results in an irremovable uncertainty in quantum ensembles. This may be illustrated as follows. Suppose that we have a classical ensemble with a random spread in some mechanical quantity L, the *variance* of this being

$$\overline{\Delta L^2} = \overline{(L - \bar{L})^2} > 0 \tag{2}$$

the mean represented by the bar being taken over the ensemble and L being the mean of L. This quantity is measured, and we select specimens having the same L, say equal to L', from which we construct a new ensemble, in which $\overline{\Delta L^2} = 0$. Speaking generally, this process of selection has no effect on the state of the system as regards other dynamic variables, e.g. the variable M. Proceeding as above, we produce an ensemble in which $\overline{\Delta L^2}=0$, $\overline{\Delta M^2}=0,\ldots$, i.e. one in which each of the dynamic variables has a completely definite value. In particular, such an ensemble might be built up from classical systems whose temperature is absolute zero, in which case the momentum $p=0$, while the coordinate has the value $q=q_{\min}$ corresponding to minimum potential energy; appropriate choice of origin allows us to put $q=0$. All the other quantities are functions of p and q, so they too have definite values corresponding to $p=0$ and $q=0$.

Nothing comparable occurs in the region ruled by quantum mechanics. A quantum ensemble with a definite value for some mechanical quantity $L(\overline{\Delta L^2}=0)$ can never be one in which every other dynamic variable also has definite value. Let L be represented by the operator $\mathscr{L}$, with some other mechanical quantity M represented by operator $\mathscr{M}$; Schwarz's inequality, which applies[1,2] to the linear and self-conjugate operators $\mathscr{L}$ and $\mathscr{M}$, gives us that

$$\overline{\Delta \mathscr{M}^2}\cdot\overline{\Delta \mathscr{L}^2} \geqslant \tfrac{1}{4}|\bar{C}|^2 , \tag{3}$$

in which $C=\mathscr{M}\mathscr{L}-\mathscr{L}\mathscr{M}$ and the bar denotes a mean over the ensemble. This is the uncertainty principle in its most general form. In particular, if M and L represent the canonically conjugate momenta p and coordinate q, and we have for the operators $\mathscr{P}$ and $\mathscr{Q}$ that

$$\mathscr{P}\mathscr{Q} - \mathscr{Q}\mathscr{P} = i\hbar, \tag{4}$$

then (2) implies that

$$\overline{\Delta p^2}\cdot\overline{\Delta q^2} \geqslant \tfrac{1}{4}\hbar^2 , \tag{5}$$

i.e. the uncertainty relation for p and q. Also, (3) shows that only the quantities represented by commuting operators

$$\mathscr{M}\mathscr{L} - \mathscr{L}\mathscr{M} = 0, \tag{6}$$

have unrelated variances $\overline{\Delta M^2}$ and $\overline{\Delta L^2}$, and hence may be chosen as specimens of microsystems in an ensemble in such a way that the new ensemble has both variances zero. Quantities represented by non-commuting operators:

$$\mathscr{M}\mathscr{L} - \mathscr{L}\mathscr{M} \neq 0, \tag{7}$$

cannot give an ensemble with $\overline{\Delta M^2}=0$ and $\overline{\Delta L^2}=0$, apart from exceptional states of the initial ensemble such that $C\psi = 0$. Hence an ensemble with a definite value for some dynamic variable L represented by operator $\mathscr{L}$ always allows us find dynamic variables M represented by operator $\mathscr{M}$ that does not commutate with $\mathscr{L}$. This means that $\overline{\Delta M^2} \neq 0$ in this ensemble, in accordance with the general uncertainty relation (3). An attempt to eliminate the variance of M by selecting specimens having a definite value M' will give rise to a new random set in which, by virtue of (3), there will be a spread in $L(\overline{\Delta L^2}$ no longer zero). Randomness cannot be eliminated in the quantum region. This feature gives rise to an extremely basic requirement in respect to the working principle of any measuring instrument designed for measurements on a quantum ensemble. Only external measuring instruments can serve to detect ensembles and to analyse distributions in them (such instruments must stand outside the ensembles) in any study of the phenomena by statistical methods; they must be free from the elements of randomness inherent in the statistical sets examined by their use. However, a measuring instrument resembles any other body in consisting of atoms, molecules, and similar micro-formations, which perform random chaotic movements, as observed in Brownian motion. Before the era of quantum mechanics it was assumed that this random molecular motion could be frozen, but we know that this motion does not stop even at absolute zero; the random fluctuations in the positions of the microparticles persist even in an absolutely cold body. The world of microparticles shows too lively disordered motion for any part of it to be used as a measuring instrument; such a micro-instrument would be permanently subject to numerous random disturbances and so itself would need to be checked continually. We need to find an island of peace in the chaos of the microworld, in the stormy sea of microphenomena. Quantum mechanics finds this island in a macroscopic instrument.

The output of any instrument is always a macroscopic effect: deflection of a pointer in a meter, formation of droplets in a Wilson cloud chamber, blackening of grains in a photographic emulsion, and so on. Quantum theory identifies this concept of macroscopicity with that of a classical nature in the instrument; in other words, the instrument must be so constructed that its action ultimately depends solely on its macroscopic features, i.e. ones in which Planck's constant plays no part. Such an instrument is[3] maximally free from quantum randomness, so we may correctly say that quantum mechanics examines the microworld in its relation to the macroworld. Macroscopic (classical) instruments are recording systems in relation to which the state of the microsystems is determined in quantum mechanics.

Actually, such a recording system may be realized only approximately, but in this respect there is no difference from the classical position. For instance, in classical mechanics we would say "The deductions are strictly correct only for measurements with absolutely rigid measures and uniformly running clocks". If these do not in fact exist, we find some devices approximating to them such that the theoretical deductions still apply to the real situations.

Now let us make a few comments on the position of man himself as a measuring instrument. There is no doubt that here we enter very tricky ground, since in the current ground state of our knowledge we cannot analyse man with the clarity and completeness possible for a physical instrument. For example, we cannot say in respect of living matter in which cases we can ignore Planck's constant and in which we cannot. We must therefore resort to less precise evidence and criteria.

A major feature is that the sense organs respond in a smoothed fashion to microevents, i.e. in a macroscopic way. For example, we sense heat or cold, the sensation being produced by microprocesses (random motion of the molecules in the surrounding medium) but with the fluctuations averaged out. The impacts of the individual molecules are not sensed, and we cannot judge the fate of single molecules. The same may be said of the perception of light; even under the limiting conditions of Vavilov's experiments[4], the eye merely establishes the fact that the intensity fluctuates at extremely low levels. Here also the final process within the man is of the ordered macroscopic type; the observer makes a note of the observed fluctuation, and in this respect he does not differ from a particle counter that moves its pointer. We are not even able to say how suitably equipped living matter would sense the direct motion of microparticles that it was capable of perceiving. The consciousness of such a being would perform some kind of Brownian motion. Should we envy such a capacity or regret such a misfortune? No reliable answer can be given even to this question.

The chaos of microphenomena appears as disproportionately large fluctuations in the chain reaction in a nuclear reactor brought into operation at very low mean power; such fluctuations greatly exceed the mean level. This reflection of atomic randomness can well get on the nerves of operators participating in the startup, since it may seem that the process will get out of control. However, these disturbing fluctuations are smoothed out when the power level becomes high. We cannot escape the impression that human perception is better adapted to the smooth course of events in the macroscopic world.

Thus although we cannot apply to man the simple criterion of macroscopicity used in quantum mechanics, we can say that human perception smooths out the atomicity of the world and in this respect is macroscopic.

Man must be considered as a macroscopic instrument from the viewpoint of physics, but he is far from being perfect in that respect, especially because he needs an extension of the facilities provided by his senses in order to study the microworld. The extension is provided by measuring instruments, which allow microphenomena to speak in the language of macrophenomena. This has given rise to the frequently debated question of how far this approach "restricts" our perception of the microworld.

Whilst this aspect will not be considered here, because it deals with epistemology rather than physics, it is difficult to avoid a few comments.

It is an old and well known truth that the criterion of the reality of perception is experience. However, the detailed mechanism of perception may vary greatly from one living being to another. This mechanism possesses certain features common to all beings, even though it may appear paradoxical from the viewpoint of man. These common features may perhaps be best expressed in the language of cybernetics: consciousness (general knowledge) gives a command to an effector organ of the living matter (or to the machines or apparatus at its disposal), the effector produces an action, and after this action the consciousness receives a return signal, communicating to it the result of the action. Feedback of this kind stimulates the consciousness to further action. Perception of the external world is based on this feedback, whose precise form may vary greatly. It is clear that if the feedback operates incorrectly (does not reflect reality), then no organism or a whole group saddled with this deficiency could exist for any prolonged period. However, the mechanisms of "reflection" may vary greatly. For example, in the devices enabling man to see at night infrared is converted into visible light; the colour of an object may be transmitted as a colour signal (by a photochemical action in the eye), or as a sound if the light is converted to sound by a photocell and then passed to the organs of hearing.

Iolanthe, blind from birth, would have found a way of distinguishing a red rose from a white one if she had been a physicist.

These examples show that the important thing is not the precise method of perceiving the signal received from the object but that the signal should be unambiguous.

For this reason we should not belittle quantum mechanics simply because it leads us to hear the music of the microworld in the language of macroscopic effects, especially since experience in applied physics and in atomic physics research shows that this language is never an obstacle to scientific and technical progress. Could man even perceive the world in a way different from that in which he does at present (i.e. as an essentially macroscopic world) without essential alteration of his organism?

REFERENCES

1. K. V. Nikol'skii, *Quantum Processes* [in Russian], GITTL, 1940.
2. W. Heisenberg, *The Physical Principles of Quantum Theory* (Appendix).
3. D. I. Blokhintsev, *The Principles of Quantum Mechanics* [in Russian], Izd. Vysshaya Shkola, 4th edition, 1963.
4. S. I. Vavilov, *The Microstructure of Light* [in Russian], Izd. AN SSSR, 1950.

CHAPTER XII

SCHEME OF A MACROSCOPIC INSTRUMENT

Here we return to the theory, especially as regards the meaning of measurement from the theoretical point of view. We assume that we have to measure the dynamic variable L associated with a microsystem μ, which itself belongs to a quantum ensemble governed by a macrosetting M. For simplicity, the ensemble is assumed pure, so that the state of the microsystem is described by the wave function $\psi_M(q)$. The variable L is represented by the operator $\mathscr{L}$ and has the proper values $L_1, L_2, L_3,$ ..., L_n together with the corresponding proper functions $\psi_1(q)$, $\psi_2(q)$, ..., $\psi_n(q)$. Then $\psi_M(q)$ may be represented as a spectral expansion with respect to the proper functions of $\mathscr{L}$:

$$\psi_M(q) = \sum_{\eta = 1}^{\infty} C_n \psi_n(q). \tag{1}$$

The set of coefficients $C_1, C_2, ..., C_n$ should be considered as ψ_M taken in the L representation, i.e. in the frame of reference of the dynamic variable L. Here it is in order to use the notation

$$C_n = \Psi_M(L_n), \tag{2}$$

which stresses this feature explicitly, but we shall employ the simpler C_n notation, with the subscript n having two meanings: to denote the proper value L_n of L and to denote the macrosetting that produces this value.

We assume that the measurement has been performed and that we have found that L has the value L_m; after this measurement the system will then belong to a new ensemble, in which L has only the value L_m, i.e. $\overline{\Delta L^2} = 0$ (if $\overline{\Delta L^2} \neq 0$ before the measurement). The corresponding state of the system is now $\psi_m(q)$ and will relate to the new macrosetting resulting from the measurement.

Comparison of the new state $\psi_m(q)$ with the starting $\psi_M(q)$ shows that, in mathematical terms, the spectrum of (1) has been reduced to the single

70

function $\psi_m(q)$:

$$\sum_{n=1}^{\infty} C_n\psi_n(q) \to \psi_m(q). \tag{3}$$

This process is termed contraction of the wave packet or contraction of the wave function. The coefficients C_n in the expansion determine the intensities $|C_n|^2$ of the particular states $\psi_n(q)$ in the superposition of (1) and hence determine the probability that a measurement of L will give the value L_m, i.e. that measurement of L causes $\psi_M(q)$ to contract to $\psi_m(q)$ (see any textbook on quantum mechanics).

This is the purely formal side of the matter; let us now consider some examples showing how the process occurs under actual conditions.

A. ANALYSIS OF A POLARIZED BEAM

Consider a light source S (Figure 7) working into a collimator C and a polarizer $\mathscr{P}$, the result being a polarized beam propagating along the Ox axis. These parts of the macrosystem may be considered as preparing the microsystem (here a light quantum) in a definite pure state ψ_M, in which M represents the macrosetting (including the light source, collimator, polarizer, possible diaphragms, and so on; in particular, it includes the fact that the path of the beam contains no other microscopic or macroscopic bodies that could affect the propagation). The state ψ_M of the

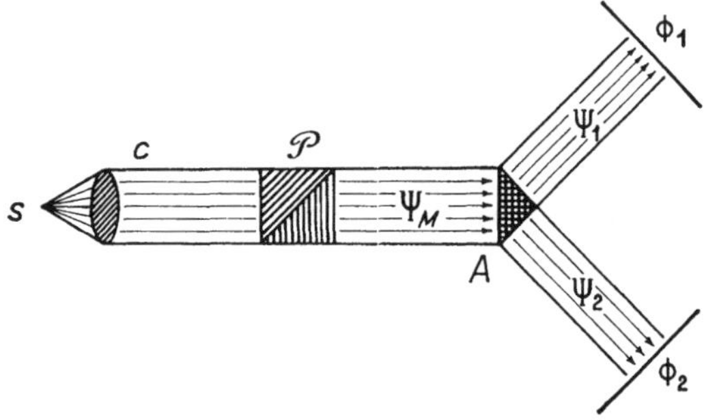

Fig. 7. Experiment with a polarized light beam. S is a light source, C is a collimator, and P is a polarizer, the last two together forming the preparative part of the equipment. This gives the ensemble ψ_M. A is an analyser which produces the two beams ψ_1 and ψ_2, while ϕ_1 and ϕ_2 are photon detectors. Thus $S + C + P = M$, $A + \phi_1 + \phi_2 = I$ (instrument).

linearly polarized light may be represented as the superposition of two mutually perpendicular states of polarization, say ψ_1 and ψ_2:

$$\psi_M = C_1\psi_1 + C_2\psi_2 \tag{4}$$

in which $C_1 = \cos\vartheta$ and $C_2 = \sin\vartheta$, with ϑ being the angle between the polarization direction of the beam M and the polarization direction in state 1. The spectral expansion of (3) is effected in practice by placing at some distance from $\mathscr{P}$ an analyser A extracting light beams having states of polarization 1 and 2 lying in different directions. These beams fall on the detectors ϕ_1 and ϕ_2, which respond to the arrival of a light quantum in one of the two possible directions, ψ_1 or ψ_2. The detector may be a photomultiplier, which is in principle capable of recording a single photon to give rise to an electron avalanche (a macroscopically recorded electrical pulse). The channel in which the pulse occurs will show whether the photon had a polarization of 1 or 2. The numbers N_1 and N_2 of the photons in the two channels are given by (3) as follows for the case of measurements repeated many times:

$$\frac{N_1}{N_2} = \frac{|C_1|^2}{|C_2|^2} = \cot^2 v. \tag{5}$$

This system is termed a measuring instrument; it has two major parts, whose purposes differ: the analyser, which performs the spectral decomposition of (4), and the detectors ϕ_1 and ϕ_2 (here photomultipliers). It is clear that the fluxes ψ_1 and ψ_2 must be laterally bounded in order to allow the detectors to function properly, otherwise ϕ_1 and ϕ_2 would confuse the assignment of a quantum to a particular beam.

It is clear that the entire device forms part of the macrosetting; the possibility of dividing the latter into a "preparative" part (usually denoted by M) and a "measuring" part is due to the spatial separation of M from $A + \phi_1 + \phi_2$.

B. MOMENTUM MEASUREMENT

Figure 8 shows schematically another experiment, in which we have a particle source S (the particles may be exposed to external electromagnetic fields ε) and collimating diaphragms C, which collimate the beam of free particles leaving the region of action of ε.

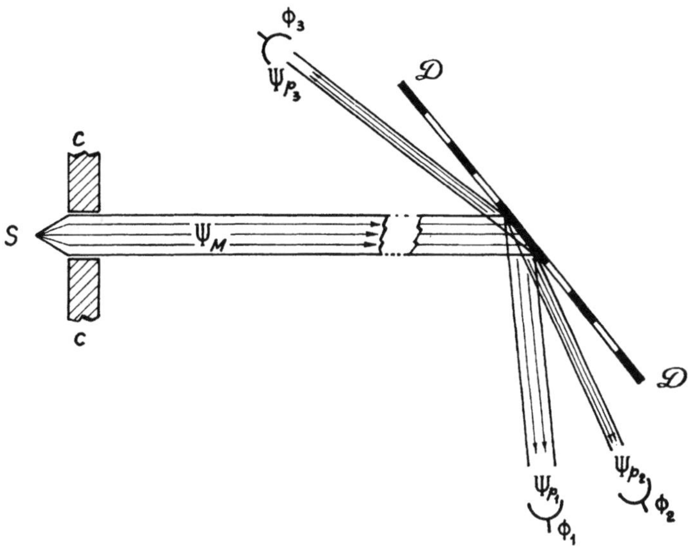

Fig. 8.   Momentum measurement with a diffraction grating D; S is particle source, and C is the diaphragm. The preparative part is $M = S + C$, while D splits ψ_M into a spectrum $\psi_1, \psi_2, \psi_3, \ldots$; $\phi_1, \phi_2, \phi_3, \ldots$ are particle detectors. $I = D + \phi_1 + \phi_2 + \ldots$.

At the exit from the collimator we have a transversely bounded electron beam with generally undefined momentum p. There are two possible causes for this uncertainty: the source may not provide particles of identical momentum (in which case we have a mixed ensemble; we assume that this cause is unimportant, the resulting spread in p thus being small), and the particles are exposed to the external varying fields ε. In the latter case the ensemble may be pure but may have a spread in momentum. This is the case that we shall consider, on the assumption that the pure ensemble is described by the wave function $\psi_M(x, y)$, which is resolved spectrally with respect to the proper functions of the momentum operator

$$\psi_p = (x) = \exp\left(ipx/\hbar\right)/(2\pi h)^{\frac{1}{2}}$$

as follows:

$$\psi\ (x, y) = \sum_p a_p(y)\, \psi_p(x), \qquad (6)$$

the lateral restriction of the beam being incorporated by assuming that the coefficients $a_p(y)$ are slowly varying functions of the y-coordinate and vanish for $|y| \gg d/2$, in which d is the beam width. This whole part of the macrosetting may be considered as the "preparatory" starting ensemble ψ_m.

The beam falls on the diffraction grating D, which produces a spatial separation of beams differing in momentum, i.e. it resolves the primary wave $\psi_M(x, y)$ into a spectrum.

This grating also alters the beam direction and, in essence, alters the initial ensemble: the initial state $\psi_M(x, y)$ is converted by D into a mixture of the primary wave $\psi_M(x, y)$ and scattered waves $b_p(x, y) \psi_{p_s}(x, y)$, which are shown as separate beams $\psi_{p_s}(x, y)$ in the figure:

$$\psi_M(x, y) \rightarrow \psi'_M(x, y) = \psi_M(x, y) + \sum_s b_s(x, y) \psi_{p_s}(x, y), \qquad (7)$$

in which the sum is taken with respect to the beams $\psi_{p_s}(x, y)$ as scattered by the grating. The spatial separation of the beams causes the coefficients for the various p_s to satisfy

$$b_s(x, y) \cdot b_{s'}(x, y) \cong 0. \qquad (8)$$

This spectral decomposition may be performed far from the "preparative" section, in which case the macrosetting may be divided into a "preparative" part and an "instrument" (here the grating plus detectors $\phi_1, \phi_2, \ldots$) that detects the particles in a given beam after scattering.

The localization of a particle in a particular detector allows us to evaluate the momentum before the grating interfered with it, i.e. before "the instrument affected the state of the object". The detector in this example may be a counter in which an electron triggers an electrical discharge i.e. which produces a macroscopic effect. Alternatively, the detector might be a photographic plate, where the electron initiates a chemical reaction in a grain of the emulsion and so causes the macroscopic effect known as a latent image.

This example shows that a diffraction grating that produces spatial separation of beams differing in momentum also destroys the interference between these, while a particle that is led to trigger a detector in a particular channel indicates its former momentum (i.e. its momentum before scattering). The physical essence of the contraction of the wave packet is thus that the microparticle alters the state of the macroscopic instrument.

74

C. DETERMINATION OF THE QUANTUM STATE OF AN ATOM

Consider a beam of atoms emerging along the Ox axis from the "black box" S (Figure 9). Neglect the detailed structure of the box and merely note that the width of the beam ΔZ_0 produced by the collimator diaphragm D is such that the momentum p_z along the Oz axis (perpendicular to the Ox axis) has the following uncertainty:

$$\Delta p_z > \frac{\hbar}{2 \cdot \Delta Z_0}. \tag{9}$$

The momentum along Ox may be taken as being defined with unlimited precision as $p_x Mv$, in which M is the mass of the atom, which will be taken as fairly large. We also shall assume that the atoms escaping from the black box may differ in their internal states, which are described by the wave functions $\psi_n(x)$, where x indicates the internal coordinates of the atom, e.g. the coordinates of its electrons. The energies of these states are E_n. Moreover, the electric or magnetic moment of the atom may vary with the state, in which case an inhomogeneous electric or magnetic field allows us to divide the beam into beams in each of which the atoms are

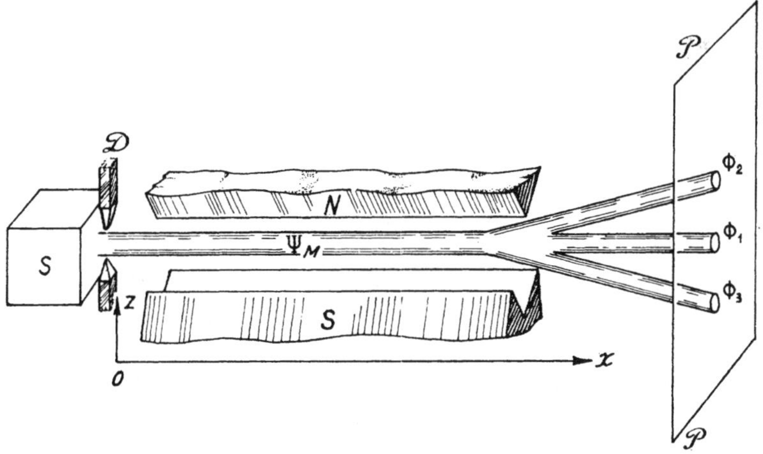

Fig. 9. Measurement of the internal energy of an atom (Stern-Gerlach experiment). S is a "black box", D is a diaphragm, $M = S + D$ prepares ψ_M. NS is a magnet (the analyser), P is a plate (detector), $I = NS + P$. $\phi_1, \phi_2, \phi_3, \ldots$ are the separated beams.

in some definite internal state $\psi_n(x)$. This division occurs for example in the Stern-Gerlach experiment on the magnetic moment of the atom (here we note that what is measured is[2,3] actually the energy of the atom in the external magnetic field H). The internal energy E_k of the atom becomes dependent on the position in space when the atom enters this external inhomogeneous field:

$$E_n \rightarrow E_n(X, Y, Z), \tag{10}$$

in which X, Y, Z denote the coordinates of the centre of gravity of the atom; in particular, since the states n have different magnetic moments $\mathcal{M}_n$ we have

$$E_n(X, Y, Z) = E_n - \mathcal{M}_n H, \tag{11}$$

in which $\mathcal{M}_n$ is the projection of the atomic magnetic moment on the direction of H, this being a function of X, Y, Z. The problem may be made more concrete by assuming this to be directed along the Oz axis and dependent on Z. The projection $\mathcal{M}_n$ equals the Bohr magneton $e\hbar/2\mu c$ multiplied by some numerical factor g, which depends on the structure of the atom (the Landé factor[2]). A small H does not alter the internal structure of the atom, so the wave functions $\psi_n(x)$ in this approximation may be taken as independent of the position of the atom in space, i.e. of X, Y, Z. These functions form a complete set, so the wave function $\Psi(X, Y, Z, x)$ for the atom as a whole is dependent on the internal coordinates x and on the coordinates X, Y, Z of the centre of gravity of the atom; the latter function may thus be expanded as a series in $\psi_n(x)$:

$$\Psi_M(X, Y, Z, x) = \sum_n \Psi_n(X, Y, Z) \, \psi_n(x). \tag{12}$$

By representing the state of the atoms in our ensemble as a superposition of the states $\psi_n(x)$ we explicitly assume that the "black box" prepares a pure ensemble Ψ_M, but with the energy E_n uncertain $(\overline{\Delta E^2} \neq 0)$. This "black box" is the preparative part of the macrosetting. The measuring instrument consists of the magnet NS, which produces the inhomogeneous field H (itself dependent on Z) along the OZ axis, together with the detectors $\phi_1, \phi_2, \ldots$; as we shall see later, the atom itself here forms part of the measuring instrument. $\Psi(X, Y, Z, x)$ must satisfy Schrödinger's equation:

$$i\hbar \frac{\partial \Psi}{\partial t} = \mathscr{H}\Psi, \tag{13}$$

$$\mathscr{H}(X, Y, Z, x) = T(X, Y, Z) + \mathscr{H}_0(x, X, Y, Z),$$

in which $\mathscr{H}_0(x, X, Y, Z)$ is the hamiltonian describing the internal motion of the atom. In the presence of an external field we have

$$\mathscr{H}_0(x, X, Y, Z)\,\psi_n(x) = E_n(X, Y, Z)\cdot\psi_n(x), \tag{14}$$

in which we have used the assumption that the external field does not deform the atom, i.e. that the $\psi_n(x)$ remain unchanged. Further, $T(X, Y, Z)$ is the hamiltonian for the motion of the centre of gravity; there is no interaction between the atom and the external field apart from that due to the magnetic (or electric) moment of the atom, so $T(X, Y, Z)$ is simply the hamiltonian for the kinetic energy of the atom as a whole.

Substituting (12) into (13), multiplying the result by one of the $\psi_m^*(x)$ functions, and integrating with respect to the internal variables (x), we obtain from the condition that the functions are orthogonal,

$$\int \psi_m^*(x)\,\psi_n(x)\,\mathrm{d}x = \delta_{nm} \tag{15}$$

a system of equations for $\Psi_m(X, Y, Z)$:

$$i\hbar \frac{\partial \Psi_m}{\partial t} = [T + E_m(X, Y, Z)]\,\Psi_m. \tag{16}$$

Here the energy $E_m(X, Y, Z)$ of the atom plays the part of the potential energy for the motion of the atom as a whole. H is macroscopic in character, so it changes smoothly on the scale of the wavelength $\lambda = \hbar/Mv$ of the atom: $\lambda\partial H/\partial Z \ll H$. This condition is simply the condition of applicability of classical mechanics to the motion of the atom as a whole. Therefore we put $\Psi_m(X, Y, Z)$ in the form

$$\Psi_m(X, Y, Z) = \sqrt{\rho_m(X, Y, Z)}\,\exp\left[\frac{i}{\hbar}\,S_m(X, Y, Z)\right], \tag{17}$$

in which $S_m(X, Y, Z)$ is the classical action function, while $\rho_m(X, Y, Z)$ is the density of the atoms in (X, Y, Z) space. We substitute (17) into (16) and neglect higher powers of $\hbar$; separating the real and imaginary parts, we have

$$\frac{\partial S_m}{\partial t} = \frac{1}{2M}(\nabla S_m)^2 + E_m(X, Y, Z) \cdot S_m, \tag{18}$$

$$\frac{\partial \rho_m}{\partial t} - \frac{1}{M}(\rho_m \nabla S_m) = 0. \tag{18'}$$

The first of these is the Hamilton-Jacobi equation for the action function S, while the second is the equation of continuity for the density ρ_m in beam m; $-(1/M)\nabla S_m$ is simply the velocity v_m of the atoms in the beam; thus this equation states that the particles will move in such a way that their flux is constant in any cross-section of the tube formed by the trajectories of the particles. We do not need to solve these equations; it is simpler, in view of the classical character of the motion, to use Newton's equations directly:

$$M\frac{d^2 X_m}{dt^2} = 0, \quad M\frac{d^2 Y_m}{dt^2} = 0, \quad M\frac{d^2 Z_m}{dt^2} = -\frac{\partial E_m}{\partial Z}. \tag{19}$$

These equations give

$$X_m = vt + X_0, \quad Y_m = Y_0, \quad Z_m = \frac{1}{2M}\frac{\partial E_m}{\partial Z}t^2 + Z_0, \tag{19'}$$

in which X_0, Y_0, Z_0 are the initial values of the coordinates X, Y, Z; v is the velocity of the atom along the OX axis. It must be remembered that this solution is approximate, for the atoms that pass through D will not move along the classical loci; the beam will spread out. This quantum effect may be allowed for by taking a further step in the approximate solution of (16) by incorporating terms containing $\hbar$ to the first power. We shall not do this here, but merely limit ourselves to estimates. The uncertainty relation (9) implies that the beam width will increase in the OZ direction.

Thus, from (9) it follows that

$$v_z > \frac{\hbar}{2M\Delta Z_0},$$

and so that

$$\Delta Z_t = v_z t > \frac{\hbar}{2M\Delta Z_0} \cdot t$$

78

Beams relating to different internal states n and m will be separated provided that

$$\left|\frac{\partial E_m}{\partial Z} - \frac{\partial E_n}{\partial Z}\right| \frac{t^2}{2M} > \frac{\hbar t}{2M \Delta Z_0} \quad \text{on}$$

$$\left|\frac{\partial E_m}{\partial Z} - \frac{\partial E_n}{\partial Z}\right| \Delta Z_0 \cdot t > \hbar. \tag{20}$$

Since E_n varies only slightly with Z, we may put

$$|E_n - E_m|\, t > \hbar, \tag{21}$$

which is the usual relation between the energy E of the state under measurement and the duration t of that measurement; this relation should not[4] be confused with the uncertainty relation for p and q.

Consider now the operation of the detectors $\phi_1, \phi_2, \ldots$ receiving the beams.

In what follows it is convenient to assume that the functions $\Psi_n(X, Y, Z)$ for the motion of the centre of gravity are normalized to 1 throughout the (X, Y, Z) space; for this purpose it is sufficient to put

$$\Psi_n(X, Y, Z) = C_n \phi_n(X, Y, Z),$$

whereupon the superposition of (12) becomes

$$\Psi_M(X, Y, Z) = \sum_n C_n \phi_n(X, Y, Z)\, \psi_n(x). \tag{22}$$

This implies that the probability that the atom lies near the point X, Y, Z is

$$W(X, Y, Z)\, dX\, dY\, dZ = \int |\Psi_u(X, Y, Z, x)|\, dx$$

$$= \sum_n |C_n|^2\, |\phi_n(X, Y, Z)|^2\, dX\, dY\, dZ. \tag{23}$$

We shall consider now the probability that the atom is in beam m, which is equivalent to its having an energy E_m; (23) shows that this probability is the integral of $W(X, Y, Z)$ over the volume V_m of beam m. If the beams for the various m do not overlap, which is possible if (21) is obeyed, the integral is

$$W_m = \sum |C_n|^2 \int_{V_m} |\phi_n(X, Y, Z)|^2\, dX\, dY\, dZ = |C_m|^2, \tag{24}$$

in complete agreement with the interpretation of the coefficients C_m as indicating the relative participation of the states ψ_m in the initial ensemble Ψ_M of (22).

The detectors record the occurrence of the atom in a particular beam. The detector may be a "cold" plate placed in the path of the beam (plates ϕ_1, ϕ_2, ... in Figure 9). An atom falling on such a plate gives up its energy and is absorbed.

It would be simple to show that the flux at the plate in each beam is proportional to the probability W_m. Hence the concentration of the deposited atoms on the plate may be measured to define $|C_m|^2$. This instrument is "classical" and macroscopic not only because the fields are macroscopic but also because of the classical character of the motion of the centre of gravity of the atom itself. In this example the atom is used as part of the macroscopic devices intended to elucidate the internal state of the atom. An atom is a relatively heavy system and does not produce marked diffraction effects; it moves along a nearly classical path and sticks to the cold plate at a well-defined point, so that the sticking of the atom in this example is itself a macroscopic effect, although not on a very large scale.

In this example we again find the previous characteristic features of the measurement process: macroscopic fields and macroscopic motion of the centre of gravity disrupt the interference between the different states $\psi_n(x)$ forming the initial quantum ensemble $\Psi_M(X, Y, Z, x)$. Sticking of a heavy atom at a definite spot on the plate serves as the basis for "contracting" the initial wave function Ψ_M of (22) to one of the particular states $\psi_n(x)$: $\Psi_M \rightarrow \psi_n$.

REFERENCES

1. P. A. M. Dirac, *The Principles of Quantum Mechanics*, 3rd ed., 1947.
2. D. I. Blokhintsev, *The Principles of Quantum Mechanics* [in Russian], Izd. Vysshaya Shkola, 4th ed., 1963.
3. W. Pauli, *The General Principles of Wave Mechanics*.
4. L. I. Mandel'shtam, *Collected Works* [in Russian], Vol. **3**, p. 397, Izd. AN SSSR, **1950**.

CHAPTER XIII

THE THEORY OF MEASUREMENT

The essence of a measuring instrument I may be expressed as

$$I = A + D$$

in which A denotes an analyser and D a detector.

Previously we have considered mainly the theory of the analyser and left the description of the detector on one side. In particular, we have described in detail the action of an inhomogeneous magnetic field on the motion of an atom, which in the example given (Chapter XII) served as the classical mechanism which performs the spectral resolution of the initial ensemble. We have not dealt in detail with the theory of the diffraction grating that resolves the ensemble with respect to p_s simply because that theory is well known and is largely the theory of Bragg reflection. For the same reason we did not go into the mathematical theory of the light polarizer in the example on the resolution of light waves into spatially separate polarized beams. That theory would, strictly speaking, be even more out of place in this book, for it relates to the theory of electromagnetic fields rather than to quantum mechanics proper.

On the other hand, we avoided mathematical description of the operation of the detector for quite opposite reasons: the detailed mode of action of the detector is usually very complex and is best considered as a separate topic; this topic forms the subject of this section. The complexity is probably the reason why the theory of the detector is commonly not presented in textbooks on quantum mechanics, but this major omission is the source of many misunderstandings and incorrect conceptions.

The mathematical theory of measurement is as follows. The entire macroscopic setting $\mathfrak{M}$ may be put as the sum of parts:

$$\mathfrak{M} = \mathscr{M} + A + D,$$

in which $\mathscr{M}$ is the part of the macroscopic setting that governs the state of the initial ensemble. To avoid a complicated argument, let us assume that the initial ensemble arises in the pure state described by the wave function $\Psi_{\mathcal{M}}(x)$ (here x denotes the set of dynamic variables describing the microsystem μ). A denotes the part of that setting which acts as the analyser, which performs the spectral decomposition of $\Psi_M(x)$ into the pure states of the relevant dynamic variable, here taken as L.

The last is assumed to have a discrete spectrum of proper values $L_1, L_2, \ldots, L_n$ together with the corresponding pure states $\psi_1(x), \psi_2(x), \ldots, \psi_n(x)$ for the microsystem. We than have the part of the macrosetting containing the detector D, which essentially changes its state in response to a microparticle. This change is a macroscopic phenomenon. Let Q be the dynamic variables of the detector, which are required to describe the states and changes in the detector. Since the detector is a macroscopic device (in particular, it may be assigned a temperature θ) it is best to describe it via a density matrix $\rho_D(Q, Q')$ rather than via a wave function $\Psi(Q)$. In this connection it is also more convenient to describe the microsystem μ that interacts with the detector via a density matrix $\rho_\mu(x, x')$ rather than via a wave function $\Psi_M(x)$. The interaction means that the two matrices must be combined into one common density matrix, which now is a function of time:

$$\rho_{D+\mu} = \rho_{D+\mu}(Q, x; Q', x', t). \tag{1}$$

This matrix obeys the equation of motion

$$\frac{\partial \rho_{D+\mu}}{\partial t} + [\mathscr{H}, \rho_{D+\mu}] = 0, \tag{2}$$

$$\mathscr{H} = \mathscr{H}_D(Q) + \mathscr{H}_\mu(x) + W_{D\mu}(Q, x),$$

in which $\mathscr{H}_\mu(x)$ is the hamiltonian of μ, $\mathscr{H}(Q)$ is the same for D, and $W_{D\mu}(Q, x)$ is the operator that describes the interaction. This equation is easily written down but is very difficult to solve in cases of interest to us. It is sufficient to recall that a detector is usually a very complex macroscopic device, such as the sensitized grains in a photographic emulsion, the supercooled vapour in a cloud chamber, the electron avalanche in a Geiger counter, etc. All the same, we shall consider later some schematic examples of the operation of conventional detectors, while at present we shall deal with the operation formally on the basis of the common matrix

$\rho_{D+\mu}(Q, x, Q', x')$, which without restriction of any kind may be put as

$$\rho_{D+\mu}(Q, x; Q', x') = \sum_{m, n} \psi_m^*(x)\, \rho_{mn}(Q; Q')\, \psi_n(x'), \qquad (3)$$

in which the $\psi_n(x)$ are the proper functions of the measured quantity L. This matrix is of course non-diagonal with respect to L in the general case.

A measuring instrument that carries out its function well will give rise for a time t of adequate length to a situation such that, if the Q lie in the range $Q_n' < Q < Q_n''$ (which may be the position of the needle of a macroscopic instrument at some definite point on its scale), all elements of the matrix for $t \to \infty$ obey

$$\rho_{mn}(Q, Q', t) = 0, \qquad (3')$$

apart from

$$\rho_{nn}(Q, Q', t) \neq 0, \qquad (3'')$$

if

$$Q_n' < Q, Q' < Q_n''. \qquad (3''')$$

This condition means firstly that the interference between the individual states $\psi_n(x)$ of the microsystem is destroyed, and secondly that the value $L = L_n$ of the dynamic variable under measurement corresponds uniquely to the state Q of the detector. This theory is the most general scheme for a measurement in quantum mechanics, as it includes not only the microsystem but also both parts of the macroscopic instrument (analyser and detector).

Conditions (3'), (3''), and (3''') may be considered as technical conditions specifying a good macroscopic instrument.

A. DETERMINATION OF THE INTERNAL STATE OF AN ATOM

Here we consider an example of such a determination by the method of deflection in an external field (Chapter XII), with allowance for the operation of the detector.

We must now supplement our system of atom + electron with a detector, which in this case is a "cold" plate that adsorbs the atom at its surface. Let q denote a set of dynamic variables describing this plate, e.g. the normal coordinates describing the vibration of its atoms. The plate has a temperature θ. The complete density matrix describing atom + electron +

83

detector may be put as

$$\rho(Q, x, q; Q', x', q', \theta, t) = \sum_{m,n} \Psi_m^*(\theta, t) \cdot \psi_m^*(x) \cdot \rho_{mn}(q, q', \theta, t)$$

$$\Psi_n(Q', t) \cdot \psi_n(x'), \qquad (4)$$

in which $\Psi_n(Q, t)$ and $\psi_n(x)$ have the meanings of Chapter XII; $\psi_n(x)$ describes the state of the electrons in the atom, while $\Psi_n(Q, t)$ describes the state of motion of the atom as a whole; the matrix element $\rho_{mn}(q, q', \theta, t)$ describes the state of the detector D.

The external field separates spatially the beams described by the functions $\Psi_n(Q)$, in a way such that the matrix (4) over a sufficiently long period t goes over to

$$\rho(Q, x, q; Q', x', q', \theta, t)_{t \to \infty} = \sum_n |\Psi_n(Q)^2| \cdot |\psi_n(x)|^2 \, \rho_{nn}(q, q', \theta, t). \qquad (5)$$

In this matrix $\Psi_n(Q)$ has the form

$$\Psi_n(Q) = \exp\left[-\frac{(Q - Q_n)^2}{2a^2} \right] \cdot H\left(\frac{Q - Q_n}{a} \right), \qquad (6)$$

and describes the state of an atom adsorbed on the plate near the point $Q = Q_n$. This atom vibrates about that position with an amplitude $a = (h/2M\omega_0)^{\frac{1}{2}}$, in which M is the mass of the atom and ω_0 is the vibration frequency. The matrix element $\rho_{nn}(q, q', \theta, t)$ describes the excited state of the lattice of atoms in the plate that arises from the transfer of the energy of the adsorbed atom to the atoms on the plate. The fact that the function describing the n-th beam has become $\Psi_n(Q)$ means that the "pointer" of the instrument (here the heavy atom) has taken up a definite position on the scale of the instrument (on the "cold" plate). The general theory developed in the previous section shows that the spatial separation of the beams implies that all terms in the matrix of (5) are zero, except the n-th, for a given coordinate of the atom adsorbed around $Q \simeq Q_n \pm a$.

Adequate performance in our detector requires that each atom falling on the plate from the beam should be adsorbed there. A necessary condition for such adsorption is that the temperature θ of the plate should be low.

Under these conditions the plate + atom system is unstable in the sense that the plate is heated by the energy received from the incident atom (the entropy of the system is increased). The reverse process is improbable, since it would be accompanied by a reduction in the entropy.

B. MOMENTUM OF A MICROPARTICLE DETERMINED BY INTERACTION WITH A MACROSCOPIC BODY

Consider now a simple but somewhat formal scheme[1] for the determination of the momentum k of a microparticle μ from the latter's interaction with a macroscopic body.

It is clear that this body must be in a state of unstable (or nearly unstable) equilibrium, as otherwise the microparticle would be unable to "disturb" it. As our body we take a sphere of mass M whose centre of gravity has a coordinate Q and whose potential energy $U(Q)$ is of the form shown in Figure 8. The sphere thus lies at the peak in U, and its relative stability there is due to a minor relative minimum in $U(Q)$. Acquirement of a small amount of energy $\Delta E > U_0 - E_0$ will cause the sphere to roll down the flank. $U(Q)$ thus takes the form of a high volcano with a small crater (Figure 10). This sphere is also a detector, determining the direction of the momentum of the microparticle, which may knock the sphere to right or left.

We have greatly simplified the problem by assigning to the sphere only the one degree of freedom Q, so it will be more convenient to use wave functions (rather than density matrices) to deal with the problem.

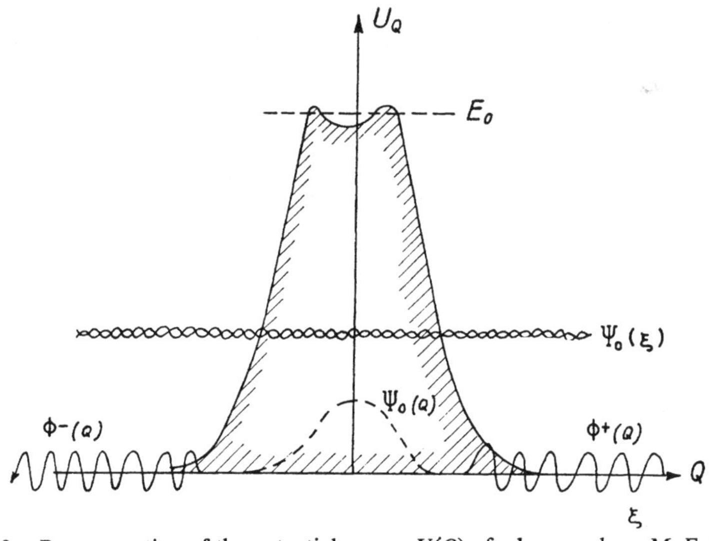

Fig. 10. Representation of the potential energy $U(Q)$ of a heavy sphere M, E_0 being the initial state. The wave function $\psi_0(Q)$, $\phi^{\mp}(Q)$ and $\psi_0(\xi)$ are also shown.

We assume that at the start ($t=0$) the microparticle μ is described by

$$\psi_0(\xi) = A^+ e^{ik\xi} + A^- e^{-ik\xi}, \tag{7}$$

in which ξ is the coordinate of the microparticle (momentum k). This means that we assume a pure state, but with an undefined momentum $\pm k$. The task of our instrument is to determine the sign of the momentum (direction of motion of the particle).

The wave function of the sphere M at $t=0$ is

$$\Psi_0(Q) = \frac{1}{\sqrt{\pi}} \exp\left(-\frac{Q^2}{2a^2}\right), \tag{8}$$

in which $a=(\hbar/M\omega_0)^{\frac{1}{2}}$, ω_0 being the vibrational frequency of the sphere within the crater.

The wave function of the whole system $(\mu+M)$ at $t=0$ is then

$$\phi(Q,\xi,0) \approx \phi_0(Q,\xi) = \Psi_0(Q)\,\psi_0(\xi). \tag{9}$$

The hamiltonian for this system is

$$\mathscr{H}(Q,\xi) = -\frac{\hbar^2}{2M}\frac{\partial^2}{\partial Q^2} + U(Q) - \frac{\hbar^2}{2\mu}\frac{\partial^2}{\partial \xi^2} + W(Q,\xi), \tag{10}$$

in which $W(Q,\xi)$ is the energy of the interaction between M and μ, the latter being considered as free and the former having the potential energy $U(Q)$. For simplicity we assume $W(Q,\xi)$ to have the form

$$W(Q,\xi) = g\cdot\delta(Q-\xi) \tag{11}$$

and that the wave function $\phi(Q,\xi,t)$ at any time t obeys the equation

$$i\hbar\frac{\partial\phi}{\partial t} = \mathscr{H}(Q,\xi)\,\phi. \tag{12}$$

This function is sought in the form

$$\phi(Q,\xi,t) = \phi_0(Q,\xi) + \phi^+(Q,\xi,t) + \phi^-(Q,\xi,t). \tag{13}$$

On the assumption that the coupling constant g is small, we find ϕ^+ and

ϕ^- in the first approximation of perturbation theory as

$$\phi^+ (Q, \xi, t) = \int U^+_{p'k'}(t)\, \Psi_{p'}(Q)\, e^{ik'\xi}\, e^{i(\omega_{p'} + \omega_{k'})t}\, dp'\, dk'$$

$$= e^{i(\omega_0 + \omega_k)t} \int U^+_{p'k'}(t)\, \Psi_{p'}(Q)\, e^{ik'\xi}\, e^{-i\Omega t}\, dp'\, dk', \tag{14}$$

$$\Omega = \omega_0 + \omega_k - \omega_{p'} - \omega_{k'} = \frac{z}{t}. \tag{15}$$

Here $\hbar\omega_0 = E_0$ is the energy of the sphere in its initial state, $E_{\mathbf{p}'} = \hbar\omega_{p'}$ the energy in the final state, P' being the momentum of the sphere after transition to the excited state; $\hbar\omega_k = \varepsilon_k$ and $\hbar\omega_{k'} = \varepsilon_{k'}$ are the energies of the particle before and after interaction with the sphere.

The form of $\phi^- (Q, \xi, t)$ is analogous.

Next, we integrate (12) after substituting from (13) and using (14) to get

$$U^+_{p'k'}(t) = \frac{1}{\hbar} \frac{(e^{i\Omega t} - 1)}{\Omega} \cdot U^+_{p'k',\,0k'}, \tag{16}$$

in which the matrix element on the right is

$$U^+_{p'k',\,0k} = gA^+ \int \Psi^*_{p'}(Q)\, e^{-ik'\xi}\, \delta(Q - \xi)\, \Psi_0(Q)\, e^{+ik\xi}\, dQ\, d\xi. \tag{17}$$

The function $\Psi_{p'}(Q)$ for the sphere in the excited state may be put in the semiclassical approximation as

$$\Psi_{p'}(Q) \sim N_{p'} \exp\!\left(\frac{i}{\hbar}\, S_{p'}(Q)\right), \tag{18}$$

in which $N_{p'}$ is a normalization factor and $S_{p'}(Q)$ is the action function, equals approximately to $p'Q$. For this reason the integral in (17) equals the Fourier transform $\tilde{\Psi}_0(\alpha)$ of $\Psi_0(Q)$, with $\alpha = p' + k' - k$. Hence

$$\phi^+ (Q, \xi, t) = e^{i(\omega_0 + \omega_k)t}\, \frac{gA^+}{\hbar} \int N^*_{p'}\, \tilde{\Psi}_0(p' + k' - k)\, \frac{(1 - e^{-i\Omega t})}{\Omega}$$
$$\Psi_{p'}(Q)\, e^{ik'\xi}\, dp'\, dk'. \tag{19}$$

Now let P be the momentum of the sphere such as to correspond to conservation of energy during the interaction; then (15) shows that this will be so for $z/t=0$ and

$$\omega_p = \frac{E_p}{\hbar} = \omega_0 + \omega_k - \omega_{k'}, \quad E_p = \frac{P^2}{2M} + \text{const},$$

so that

$$\frac{P'^2}{2M} - \frac{P^2}{2M} = -\frac{Z}{t}. \tag{20}$$

Hence we find that

$$P' - P = -\frac{z}{vt}, \quad dP' = -\frac{dz}{vt},$$

in which $v=P/M$ is the velocity of the sphere. Further, for $z \simeq 0$

$$\frac{k'^2}{2\mu} - \frac{k^2}{2\mu} = \omega_0 - \frac{P'^2}{2M} \tag{21}$$

or

$$(k' - k)(k' + k) = 2\mu\omega_0 - \frac{\mu}{M} P'^2. \tag{21'}$$

If $\Psi_0(Q)$ is not too sharp (the amplitude is not very small, which will occur if the crater is not very deep), the Fourier transform $\tilde{\Psi}_0(p'+k'-k)$ of this will differ appreciably from zero only for

$$p' + k' - k \cong 0. \tag{22}$$

Then (21) and (22) with $M \rightarrow \infty$ give us that

$$k' = -k, \tag{23}$$

$$P = 2k, \tag{24}$$

which is as would be expected for a light particle colliding with a heavy but weakly bound sphere; the light microparticle is reflected elastically, the energy transfer being small (vanishingly small for $M \rightarrow \infty$).

Now let

$$p' + k' - k = P + k' - k - \frac{z}{vt} = q$$

in such a way that

$$dk' = dq, \ k' = q - P + k + \frac{z}{vt}$$

We insert the new integration variables, q and z, into (19) to obtain

$$\phi^+(Q,\,\xi,\,t) = \frac{gA^+}{\hbar}\,\exp\left[i(\omega_0 + \omega_k)\right]\frac{t\,|N_p|^2}{v}$$

$$\Psi_0(\xi)\,e^{iPQ}\,e^{i(k-P)\xi}\,I^+\left(\frac{Q-\xi}{vt}\right), \qquad (25)$$

in which

$$I^+\left(\frac{Q-\xi}{vt}\right) = \int\limits_{-\infty}^{+\infty}\frac{1-e^{-iz}}{z}\cdot\exp\left[-\frac{iz}{vt}(Q-\xi)\right]\cdot dz. \qquad (26)$$

If we denote the discontinuous integral by

$$J(a) = \int\limits_{-\infty}^{+\infty}\frac{e^{iaz}}{z}\,dz = \begin{cases}\pi i, & a > 0,\\ -\pi i, & a < 0,\end{cases} \qquad (27)$$

then

$$I^+\left(\frac{Q-\xi}{vt}\right) = J\left(\frac{\xi-Q}{vt}\right) - J\left(\frac{vt+\xi-Q}{vt}\right). \qquad (28)$$

We note that $vt>0$ for Φ^+, because then $P=2k>0$; (28) thus implies that $I^+ = -2\lambda i$ for $vt>Q-\xi$ and $Q-\xi>0$, otherwise $I^+ = 0$. Now ϕ^+ contains the factor $\Psi_0(\xi)$, so that only small values of $|\xi|\gtrsim a$ are important, and the result implies that $\phi^+(Q,\,\xi,\,t)$ differs from zero for $t\to\infty$ only in the region $0<Q<+\infty$, i.e. to the right of the rim of the crater. This corresponds to the receipt of positive momentum $P=2k$ from the microparticle.

We calculate $\phi^-(Q,\,\xi,\,t)$ in precisely the same way; here $P<0$, $v<0$, and I^+ is replaced by

$$I^-\left(\frac{Q-\xi}{vt}\right) = J\left(\frac{Q-\xi}{vt}\right) - J\left(\frac{vt-Q+\xi}{vt}\right), \qquad (28')$$

which differs from zero for $vt<Q-\xi<0$. In this case the sphere is ejected from the crater to the left.

The density matrix for our case is then

$$
\begin{aligned}
\rho(Q, \xi; Q', \xi', t) = \phi^*(Q, \xi, t)\cdot\phi(Q', \xi', t) = \phi_0^*(Q, \xi, t)\,\phi_0(Q', \xi', t)\\
+ \phi_0^*(Q, \xi, t)\,\phi^+(Q', \xi', t) + \phi_0^*(Q, \xi, t)\,\phi^-(Q', \xi', t)\\
+ \phi^{+*}(Q, \xi, t)\,\phi_0(Q', \xi', t) + \phi^{-*}(Q, \xi, t)\,\phi_0(Q', \xi', t)\qquad(29)\\
+ \phi^{+*}(Q, \xi, t)\,\phi^-(Q', \xi', t) + \phi^{-*}(Q, \xi, t)\,\phi^+(Q', \xi', t)\\
+ \phi^{+*}(Q, \xi, t)\,\phi^+(Q', \xi', t) + \phi^{-*}(Q, \xi, t)\,\phi^+(Q', \xi', t).
\end{aligned}
$$

All terms in this matrix, except the last two, vanish for $t\to\infty$ and $|Q|$, $|Q'|>a$; that is, the terms containing ϕ_0 vanish as $\exp(-Q^2/a^2)$ or $\exp(-Q'^2/a^2)$ for $Q, Q'\to\pm\infty$, while the interference terms (which contain products of the type $\phi^{+*}\phi^-$) vanish for $t\to\infty$ on account of the properties of the functions $I^\pm([Q-\xi]/vt)$. Then for $t\to\infty$ and for $|Q|$, $|Q'|\gg a$ we have

$$
\begin{aligned}
\rho(Q, \xi; Q', \xi', t) = \phi^{+*}(Q, \xi, t)\,\phi^+(Q', \xi', t)\\
+ \phi^{-*}(Q, \xi, t)\,\phi^-(Q', \xi', t), \quad t\to\infty, \quad |Q|, |Q'| \gg a.\qquad(30)
\end{aligned}
$$

The macroscopic instrument thus destroys the interference between the states of the microparticle:

$$
A^+ e^{ik\xi} \quad \text{and} \quad A^- e^{-ik\xi},
$$

Also, for $Q, Q'\to+\infty$ we have

$$
\rho(Q, \xi; Q', \xi, t) \to \phi^{+*}(Q, \xi, t)\,\phi^+(Q', \xi', t)\qquad(31)
$$

and for $Q, Q'\to\infty$

$$
\rho(Q, \xi; Q', \xi, t) \to \phi^{-*}(Q, \xi, t)\,\phi^-(Q', \xi', t).\qquad(31')
$$

The cases of (31) and (31') correspond to finding the sphere to the right or left of the crater respectively.

We have

$$
\rho(Q, \xi; Q', \xi', t) \to 0
$$

for $Q\to+\infty$ and $Q'\to-\infty$ or $Q\to-\infty$ and $Q'\to+\infty$ (this is the case of interference between results of observations on the right and left). This is as would be expected from a "good" instrument; its "pointer" should take up one of the possible definite positions. In our case the heavy sphere acts as the pointer.

90

C. THERMODYNAMICALLY UNSTABLE DETECTOR

The following is a schematic example of such a detector for a micro-particle. The latter may be an atom with one valency electron, which thus gives the atom as a whole a magnetic moment equal to that of the electron, $M_B\boldsymbol{\sigma}$, in which M_B is the Bohr magneton and $\boldsymbol{\sigma}(\sigma_x, \sigma_y, \sigma_z)$ is the Pauli spin matrix. The wave function Ψ for the atom may be put as

$$\Psi(Q, x) = \Psi_1(Q)\,\psi_1(x) + \Psi_2(Q)\,\psi_2(x), \tag{32}$$

in which $\Psi_1(Q)$ and $\Psi_2(Q)$ are functions describing the motion of the atom as a whole; $\psi_1(x)$ and $\psi_2(x)$ are functions describing the internal state of the atom and correspond to the two possible orientations of the atomic magnetic moment.

The magnetic field H is assumed to act along the OZ axis, so ψ_1 corresponds to the moment lying parallel to the OZ axis, and ψ_2 to the moment antiparallel to the axis. This field is also inhomogeneous, and the beams differing in orientation of the magnetic moment are already spatially separate, so

$$\Psi_1(Q)\,\Psi_2(Q) = 0, \tag{33}$$

as described in Chapter XII. Thus we will suppose that the first purpose of the measuring instrument [destruction of the interference between states $\psi_1(x)$ and $\psi_2(x)$, corresponding to different orientations of the spin of the valency electron] has already been fulfilled; in other words, transmission through the inhomogeneous field leaves us merely to place in each beam the appropriate detector, which will record the entry of particles corresponding to that beam, i.e. the state of a particle ("state" in the sense of spin orientation).

Such a detector might be a large set of oscillators $S = 1, 2, ..., N$ with $N \to \infty$; to avoid complexity of notation, we can assume these to be two-dimensional (oscillating in the xy-plane). We further assume that oscillations in x in practice do not interact with those in y.

This means that the x oscillations may have a temperature θ distinct from that of the y oscillations; thus we assume that in the initial state of the detector (before the interaction with μ) the x modes are coupled to a Gibbs thermostat of temperature θ, so these x modes have that temperature also at $t = 0$, whereas the y modes are assumed to be at absolute zero at $t=0$. The detector is thus in a thermodynamically unstable state; the slightest coupling between the x and y modes will rapidly cause transfer of energy from the x modes to the y ones.[2]

The temperature rise in the y modes is the macroscopic effect that gives proof of the state of the given individual microparticle (in this case an atom).

The operation of such a detector may now be considered mathematically. The hamiltonian of the unperturbed system of oscillators is written as

$$H_0 = \sum_{s=1}^{N} H_0(x_s, y_s) - E_0, \tag{34}$$

$$H_0(x_s, y_s) = -\frac{\hbar^2}{2M}\left(\frac{\partial^2}{\partial x_s^2} + \frac{\partial^2}{\partial y_s^2}\right) + \frac{M\omega_0^2}{2}(x_s^2 + y_s^2), \tag{35}$$

in which $E_0 = \hbar\omega_0 N/2$ is the zero-point energy, M is the mass of the oscillators, and ω_0 is their natural frequency. These oscillators interact with an incident beam microparticle, the interaction energy being

$$W = \omega \sum_{s=1}^{N} \mathbf{M}_s\,\sigma_s = -i\hbar\omega \sum_{s=1}^{N} \sigma_z \frac{\partial}{\partial \varphi_s}, \tag{36}$$

in which $\mathbf{M}_s$ is the mechanical momentum of the oscillator and σ_z is the spin matrix for the optical electron in the atom. We have assumed that the spins are oriented along the Z axis in the beam, so $\mathbf{M}\sigma$ may be replaced by $M_z\sigma_z$, in which

$$M_z = -i\hbar\left(x_s \frac{\partial}{\partial y_s} - y_s \frac{\partial}{\partial x_s}\right) = -i\hbar \frac{\partial}{\partial \varphi_s},$$

and σ may be replaced simply by

$$\sigma_z = \begin{bmatrix} 1 & 0 \\ 0 & -1 \end{bmatrix}$$

obviously, we take σ_z as $+1$ for one detector and -1 for the other. We note that

$$x_s = r_s \cos\varphi_s, \quad J_s = r_s \sin\varphi_s, \tag{37}$$

$$r_s = +\sqrt{x_s^2 + y_s^2}. \tag{37'}$$

Detector D is described via a density matrix ρ taken in the (x, y) representation. By x we understand all the x coordinates of the oscillators $(x_1, x_2, ..., x_s, ..., x_n)$ and by y the corresponding coordinates $(y_1, y_2, ..., y_s, ..., y_n)$; then ρ may be put as

$$\rho = \rho(x, y; x', y', t).$$

The matrix ρ satisfies the following equation (see Chapter VI):

$$\frac{\partial \rho}{\partial t} + [H_0 + W, \rho] = 0. \tag{38}$$

It is more convenient if ρ is replaced by the matrix

$$\tilde{\rho} = \exp\left(\frac{iH_0 t}{\hbar}\right) \rho \exp\left(-\frac{iH_0 t}{\hbar}\right). \tag{39}$$

We note that $[W, H_0] = 0$, as $\tilde{W} = W$; replacing ρ by its expression in terms of $\tilde{\rho}$ we obtain

$$\frac{\partial \tilde{\rho}}{\partial t} + [W, \tilde{\rho}] = 0. \tag{40}$$

In the substitution here for the operator W from (36) we must bear in mind the rules for multiplying matrices with continuous rows and columns. For this purpose the operator W must be put in matrix form. For instance, the operator $-i\hbar\partial/\partial q$ should be replaced by

$$P_{q'q''} \equiv i\hbar \frac{\partial}{\partial q'} \delta(q' - q''). \tag{41}$$

The multiplication $P \cdot \rho$ denotes the following:

$$(P\rho)_{q'q''} = \int P_{q'q'''} \rho(q''', q'') \, dq''' = -i\hbar \frac{\partial}{\partial q'} \rho(q', q'') \tag{41'}$$

and so on. These simple rules lead the substitution for W in (40) to give a simple equation in partial derivatives:

$$\frac{\partial \tilde{\rho}}{\partial t} + \omega \sum_{s=1}^{N} \left(\frac{\partial \tilde{\rho}}{\partial \varphi_s} + \frac{\partial \tilde{\rho}}{\partial \varphi_s'}\right) = 0. \tag{42}$$

The solution of this is elementary; the general integral is

$$\tilde{\rho} = \tilde{\rho}\,(\omega t + \varphi_1,\ \omega t + \varphi_2, ...,\ \omega t + \varphi_s, ...;\ \omega t + \varphi_N,$$
$$r_1,\ r_2, ...,\ r_s, ...,\ r_N;\ \omega t + \varphi_1',\ \omega t + \varphi_2', ...,\ \omega t + \varphi_s', ...,\ \omega t + \varphi_N',$$
$$r_1',\ r_2', ...,\ r_s', ...,\ r_N'),\tag{43}$$

in which

$$r_1,\ r_2, ...,\ r_N \quad \text{and} \quad r_1',\ r_2', ...,\ r_N'$$

appear as parameters.

Consider now the initial data for this matrix. To avoid cumbrous formulae for the factors, we use as a unit of length the quantity $l = (\hbar/2M\omega_0)^{\frac{1}{2}}$, while θ is replaced by $\beta = \hbar\omega_0/\theta$. All the quantities become dimensionless in these units. For $t = 0$ the matrix is

$$\tilde{\rho}\,(x,\ y;\ x',\ y',\ 0) = \rho_\theta(x,\ x')\,\rho_0(y,\ y').$$

Making the appropriate assumption that the y modes are at absolute zero, we have

$$\rho_0(y,\ y') = C_0 \cdot \exp\left[-\tfrac{1}{2}\sum_{s=1}^{N}(y_s^2 + y_s'^2)\right]\tag{44}$$

in which C_0 is a constant of normalization, while $\exp(-y_s^2/2)$ is the wave function describing the zero-point oscillation of oscillator S along the Oy axis.

The $\rho_\theta(x,\ x')$ matrix is much more troublesome, because the x' modes are at a temperature θ. Here the state is mixed, and the weights of the individual states $\psi_n(x)$ (energies E_n) are

$$e^{-E_n/\theta} = e^{-\beta E_n}$$

so $\rho_\theta(x,\ x')$ describes an ensemble in equilibrium with a Gibbs thermostat at temperature θ and is put in the form

$$\rho_\theta(x,\ x')\,e^{\beta F(\beta)}\sum_{w} e^{-\beta E_n}\,\psi_n^*(x)\,\psi_n(x') = e^{\beta F(\beta)}\,z_\theta(x,\ x'),\tag{45}$$

in which

$$z_\theta(x,\ x') = \sum_{n} e^{-\beta E_n}\,\psi_n^*(x)\,\psi_n(x').\tag{46}$$

Here the sum is first taken over all states n having the energy E_n and then over all states differing in E_n. Direct calculation of such a sum is extremely difficult even in the case of oscillators, so we use an indirect approach based on the fact that

$$\mathscr{H}(x)\, \psi_n^*(x) = E_n \psi_n^*(x)$$

and hence

$$f(\mathscr{H})\, \psi_n^*(x) = f(E_n) \cdot \psi_n^*(x). \tag{47}$$

in which $\mathscr{H}(x)$ is the Hamilton operator for the system and $\psi_n^*(x)$ is its proper function. We can then put (46) in the form

$$Z_\theta(x, x') = \sum_n e^{-\beta \mathscr{H}(x)}\, \psi_n^*(x)\, \psi_n(x') \tag{46'}$$

and by differentiating with respect to β, we can show that the sum $Z_\theta(x, x')$ satisfies the differential equation

$$\frac{\partial Z_\theta}{\partial \beta} + \mathscr{H} Z_\theta = 0. \tag{48}$$

We must replace $\mathscr{H}(x)$ by the unperturbed operator for the x modes:

$$\mathscr{H}_0(x) = \sum_{s=1}^{N} \left(-\frac{1}{2}\frac{\partial^2}{\partial x_s^2} + \frac{1}{2} x_s^2 \right) - \frac{1}{2} N, \tag{49}$$

which is taken from (34) and (35), with allowance for the new units of length for x.

The additivity of the hamiltonian of (49) allows the variables in (48) to be separated, so (48) may be solved explicitly for x alone. In this case we have

$$\frac{\partial Z_\theta(x, x')}{\partial \beta} - \frac{1}{2}\frac{\partial^2 Z_\theta(x, x')}{\partial x^2} + \left(\frac{1}{2} x^2 - \frac{1}{2} \right) \times Z(x, x') = 0, \tag{50}$$

The solution is sought in the form

$$Z_\theta(x, x') = \exp\left(a + bx^2 + C \cdot x \cdot x' + b \cdot x'^2\right) \tag{51}$$

subject to the boundary condition

$$Z_\theta(x, x') \sim \frac{1}{\sqrt{\beta}} \exp\left(-\frac{1}{2\beta}(x - x')^2 \right), \quad \theta \to \infty \quad \beta \to 0 \tag{52}$$

which corresponds to evaporation of the oscillators for $\theta \to \infty$; (52) takes the form of a sum for particles of an ideal gas.

This conclusion cannot conform with the possibility of introducing latent parameters λ, since the latter are essentially such as would, if measured, allow us to "define more closely" the quantum variables, e.g. they would allow us to alter the variance of some dynamic variable L. For instance, suppose that in the range $J_1(\lambda)$ of the latent parameters the mean $\bar{L}$ takes the value L_1, with variance $\overline{\Delta L_1^2}$, while in region $J_2(\lambda)$ the corresponding values are L_2 and $\overline{\Delta L_2^2}$. Then $\bar{L}$ and L^2 for the entire ensemble are

$$\bar{L} = \alpha_1 L_1 + \alpha_2 L_2 , \tag{4}$$

$$\overline{\Delta L^2} = \alpha_1 \cdot \overline{\Delta L_1^2} + \overline{\alpha_2 L_2^2} + \alpha_1 (L_1 - \bar{L})^2 + \alpha_2 (L_2 - \bar{L})^2 , \tag{4'}$$

in which α_1 and α_2 are the relative weights of the regions $J_1(\lambda)$ and $J_2(\lambda)$, with

$$G_2(\lambda), \quad \alpha_1 + \alpha_2 = 1, \quad \alpha_1, \alpha_2 \geqslant 0 .$$

In other words, a pure ensemble is inhomogeneous (in the parameters λ), which is contrary to the definition.

If now we seek to define more closely these λ by dividing the entire region of possible values of λ into smaller parts

$$G_1(\lambda), G_2(\lambda), ..., G_s(\lambda), ..., G_N(\lambda)$$

with weights

$$\alpha_1, \alpha_2, ..., \alpha_s, ..., \alpha_N, \sum_1^N \alpha_s = 1$$

then we find for the limit of infinitely small regions $J_s(\lambda)$ that (4') is replaced by

$$\overline{\Delta L^2} = \sum_\lambda \alpha_\lambda (L_\lambda - \bar{L})^2 , \tag{5}$$

in which

$$\alpha_\lambda = \frac{G_s(\lambda)}{\sum\limits_{s=1}^N G_s(\lambda)}$$

and L_λ is the mean of L in $J_s(\lambda)$, this mean equalling the exact value of L in region $J_s(\lambda)$ for $N \to \infty$.

Formula (5) shows that the entire random spread in L is now due to the uncertainty in the λ; the quantum dynamic variables should take com-

in which

$$\Delta = \frac{b}{2} \sum_{s=1}^{N} [r_s^2 + r_s'^2 - 2\gamma r_s r_s' \cos(\varphi_s - \varphi_s')] - \frac{1}{2} \sum_{s=1}^{N} (r_s^2 + r_s'^2), \tag{58}$$

$$A = \frac{b}{2} \sum_{s=1}^{N} [r_s^2 \cos 2\varphi_s + r_s'^2 \cos 2\varphi_s' - 2\gamma r_s r_s' \cos(\varphi_s + \varphi_s')]$$

$$- \frac{1}{2} \sum_{s=1}^{N} (r_s^2 \cos 2\varphi_s + r_s'^2 \cos 2\varphi_s'), \tag{59}$$

$$B = - \frac{b}{2} \sum_{s=1}^{N} [r_s^2 \sin 2\varphi_s + r_s'^2 \sin 2\varphi_s' - 2\gamma r_s r_s' \sin(\varphi_s + \varphi_s')]$$

$$+ \frac{1}{2} \sum_{s=1}^{N} (r_s^2 \sin 2\varphi_s + r_s'^2 \sin 2\varphi_s'). \tag{60}$$

This rather cumbrous result should be averaged over a period $\frac{1}{2}\omega$, if we assume that the frequency characterizing the atom-detector bond is fairly high.

The observed result is then defined by the matrix

$$\overline{\tilde{\rho}(x, y; x', y', t)} = C_0 C_0^{-1} \exp \left\{ \Delta \frac{1}{\pi} \int_{-\pi/2}^{+\pi/2} [(A \cos 2\omega t + B \sin 2\omega t)] \, \mathrm{d}t \right\}. \tag{61}$$

The latter integral gives rise to a Bessel function:

$$\frac{1}{\pi} \int_{-\pi/2}^{+\pi/2} \exp[(A \cos 2\omega t + B \sin 2\omega t)] \, \mathrm{d}t = I_0(R), \tag{62}$$

in which

$$R = \sqrt{A^2 + B^2}$$

so the matrix

$$\overline{\rho(x,\, y;\, x',\, y',\, t)}$$

averaged with respect to time, is

$$\tilde{\rho}(x,\, y;\, x',\, y',\, t) = C_0 C_\theta\, e^{A} \cdot I_0(R), \tag{63}$$

$$I_0(R) = 1 + \tfrac{1}{4}R^2 + \cdots, \quad |R| \ll 1. \tag{64}$$

$$I_0(R) \cong \frac{e^{R}}{\sqrt{2\pi R}} + \cdots, \quad |R| \gg 1. \tag{64'}$$

Hence for small R we have

$$\tilde{\rho}(x,\, y;\, x',\, y',\, t) = C_0 C_\theta \cdot e^{A}. \tag{65}$$

and for large R

$$\overline{\tilde{\rho}(x,\, y;\, x',\, y',\, t)} = C_0 C_\theta\, \frac{e^{A+R}}{\sqrt{2\pi R}}. \tag{65'}$$

Remembering that $b = -\tfrac{1}{2}\beta = -\theta/2$, the $\exp A$ factor containing A arising from (58) indicates that energy is transferred between the x and y modes, the temperature falling from θ to $\theta/2$.

The result for large R also shows that the energy distribution between the x and y modes alters, but this is not so obvious as for small R.

Thus we see that a microparticle entering a thermodynamically unstable detector causes there a complete change in the energy distribution, i.e. a macroscopic effect.

These examples show that a macroscopic measuring instrument must be an unstable system (more precisely, almost unstable); only then can a microparticle alter its state, and this change of state is a macroscopic event. A microparticle is powerless to act on an instrument represented by a stable macroscopic system; it cannot "displace the pointer" from the zero position.

REFERENCES

1. D. Blokhintsev, *Voprosy filosofii* **9** (1963) 108.
2. N. S. Green, *Nuovo cimento* **9** (1958) 880.

CHAPTER XIV

THE WAVE FUNCTION AS THE OBSERVER'S NOTEBOOK

This representation (as the book in which the observer records the results of his measurements on microsystems) may not be correct, as we may see from the answer usually given in textbooks. Let us suppose that from some previous measurements it is known that the wave function $\Psi_M(x)$ represents the state of the microsystems in the ensemble.

The observer enters in his book the simple sign

$$\Psi_M(x). \tag{1}$$

This in principle allows the observer to predict the probability of the results from any possible measurement on the microsystems μ.

We assume that the observer intended to measure a quantity L having (for simplicity) only two proper values L_1 and L_2, the corresponding proper functions being $\psi_1(x)$ and $\psi_2(x)$; these might for example be the spin states of the particle. Then $\Psi_M(x)$ may be represented as the superposition of the particular states $\psi_1(x)$ and $\psi_2(x)$:

$$\Psi_M(x) = C_1\psi_1(x) + C_2\psi_2(x), \tag{2}$$

in which C_1 and C_2 are coefficients defining the relative participation of the states with $L=L_1$ and $L=L_2$. We assume that the measurement is made and indicates that $L=L_1$; the observer then assigns the particle that has passed the check point (the measuring instrument) to a new ensemble characterized by the new wave function $\psi_1(x)$. The observer takes his notebook and strikes out $\Psi_M(x)$ as out of date and unsuitable for further predictions of results and fresh measurements on the particle that has undergone a measurement. He replaces $\Psi_M(x)$ by

$$\psi_1(x). \tag{3}$$

The transition from $\Psi_M(x)$ to $\psi_1(x)$ is

$$\Psi_M(x) = C_1\psi_1(x) + C_2\psi_2(x) \to \psi_1(x) \tag{4}$$

and is a process of "contraction" of a wave packet. From this viewpoint the contraction is a direct consequence of the change in the observer's information. The wave function itself is simply a routine record of his information on the state of the ensemble of microsystems. There is nothing wrong in this very common view of the wave function and of the contraction process, and it is convenient as a formulation against which it is difficult to raise objection. In using the words "observer", "measurement", "information", and so on we merely pay respect to the physicists' common jargon, which is in no way a lucid tongue for the discussion of major philosophical and methodological topics in physics. The entire theory of measurement thereby acquires a dubious taint of subjectivism, which becomes quite insupportable if we have to answer the question whether quantum mechanics is applicable to the description of physical phenomena occurring in the absence of an observer.

The observer is not an entirely obligatory being in this world, for quantum laws would scarcely be altered by one iota if this restless observer were removed from the scene.

In one of his papers on quantum mechanics Schrödinger gave an example of superposition of states which many readers found unnerving. He considered a microsystem having two states ψ_1 and ψ_2; the first causes a Geiger counter to fire, while the second leaves it undisturbed. The firing of the counter actuates an amplifier to break a tube of prussic acid in a chamber containing a cat.[1,2,3]

Thus we have that the observer, looking into his notebook in order to predict the result of a future measurement, finds among the possible results of observation the fact of a possible interference between the states of life and death of a cat. In fact, (4) implies that

$$|\Psi_M|^2 = |C_1\psi_1|^2 + |C_2\psi_2|^2 + 2\,\text{Re}\,C_1^*C_2\psi_1\psi_2 \qquad (5)$$

and the last term indicates this astonishing possibility.

After observation of the actual event (ψ_1 or ψ_2), medicine certifies either the death of the unfortunate cat or that it is alive and well, while the wave function is condensed in that report to ψ_1 or ψ_2.

It is not difficult to see that this rather horrifying example may be made even more disturbing if we replace the cat by the observer together with medical attendants; then in the case of ψ_1 there will be no one to contract the wave function.

100

However, we must return to more realistic examples. Consider the decomposition of a radioactive atom, ψ_1 being the state before decay and ψ_2 that after decay. Theory shows that $C_1 \approx \exp(-\lambda t)$, in which t is time, with $T = 1/\lambda$ the decay period;

$$|C_2|^2 = 1 - |C_1|^2,$$

so C_2 increases with time, whereas C_1 decreases. Now imagine that we are speaking of the remote past, when no observer could have transmitted to us information on the actual fate of the radioactive atom, e.g. during the time of the dinosaurs. If the interval t separating us from that time greatly exceeds T, we can say with high probability that the atom has decayed; but it is far from being a matter of complete indifference to the atom's surroundings at what time it actually did decay.

We may recall Ray Bradbury's story of travellers into the prehistoric past who carelessly crushed a butterfly, this event then influencing the outcome of the presidential election in the USA in the year 2000.

The decay of an atom may cause a chain of events whose detailed content may be very much dependent on the exact moment of decay. On the other hand, our modern observer has not yet had the chance to contract the wave function from the Ψ containing the superposition of two possibilities: the atom has decayed (ψ_2) or is still in its initial state (ψ_1).

If the observer takes the trouble to measure the state of the atom, he will most probably find that it has decayed and is in state ψ_2; but our contemporary is very late in reaching his conclusion if $t \gg T$, because any other observer would have reached the same conclusion before him. That is, we could assert in the language of quantum mechanics that the atom has decayed, quite without reference to the observer. The event may lead to consequences dependent on the precise instant of decay and therefore cannot be related to changes in the observer's information. The observer plays no part in the events of which we speak and so must be excluded from the game.

Imagine a series of observers, of whom one is our contemporary and the others are successively earlier; in this series there will be one who was the first to record the fact of decay. This moment will be of objective significance and should be reflected in the apparatus of quantum mechanics without reference to the observer.

There is now a very obvious answer to the several paradoxes raised previously, which follows from the above analysis of the physical significance of a measurement as a macroscopic process triggered by a microsystem. Nevertheless, the importance of the observer problem leads me to consider this aspect in somewhat greater detail and to summarize the above survey of the essence of measurement.

We denote by $\mathfrak{M}$ the entire macroscopic system in which the relevant microevents occur, the system being represented as a sum

$$\mathfrak{M} = M + (A + D) + O, \tag{6}$$

in which M denotes the part of the system termed preparative and which governs the initial state Ψ_M of the ensemble of microparticles; $A + D$ denotes the part in which a microparticle produces a macroscopic event and hence gives evidence of its state, this part being nominally divided into an analyser A (which causes the macroscopic event to be dependent on the state of the microparticle) and a detector D (which changes its state in response to the microparticle). Finally we include the observer O to cover the case where he considers that he affects the behaviour of the quantum ensemble. Of course, it is usual for the experimentalist to minimize his influence on the phenomenon as one of the primary points in designing the experiment.

It will be evident that we may generally delete the O (the honourable observer) from the above sum.

The entire essence of the contraction of the wave function is expressed in the fact that the microparticle produces a macroscopic effect, and the latter is of entirely objective significance, being quite enrelated to any information the observer may have on the occurrence of the event.

The part $A + D$ may be such, or may be arranged to be such, that it serves as a measuring instrument under certain conditions, this being done deliberately. Moreover, the experimentalist can modify the part $A + D$, i.e. to carry out his measurements in another way, but subject to the restriction that he avoids thereby having any effect on the part M, as otherwise he may alter the initial ensemble. The division of $\mathfrak{M}$ into parts is only approximate, so any change in $A + D$ requires some caution.

Since the observer is interested in various measurements, we may use the following representation for a set of possible measurements in terms of symbolic forms for the macroscopic setting:

$$\mathfrak{M} = M + (A + D) + O,$$
$$\mathfrak{M}' = M + (A + D)' + O, \tag{7}$$
$$\mathfrak{M}'' = M + (A + D)'' + O,$$

in which the primes in $(A+D)'$, $(A+D)''$, etc, denote that different measuring instruments are used, whereas the part M (which defines the initial ensemble) is kept unchanged as far as possible. This simple example serves to stress once again how the concept of quantum ensembles differs from other approaches to quantum mechanics: to a given quantum ensemble defined by a macrosetting M (which may be represented mathematically by the wave function Ψ_M or the density matrix ρ_M) there belong infinitely many statistical sets of measurement results relating to the various measuring instruments

$$(A + D), \quad (A + D)', \quad (A + D)'' \ldots$$

and so on, among which there may be instruments that measure complementary dynamic variables, i.e. ones that cannot in principle be measured by the same instrument, such as space-time and energy-momentum measurements.

If we assume that the observer is justified in his decision to change part of the macrosetting, i.e. the measuring instrument $A + D$, then it is clear that objectively (i.e. without dependence on the observer) the quantum statistical collective of microsystems is specified by the fact that it remains unchanged during a replacement of one instrument by another, i.e. by the wave function Ψ_M or the density matrix ρ_M.

REFERENCES

1. E. Schrödinger, *Reports of Solvay Congress* (1928).
2. N. Bohr, *Uspekhi fiz. nauk* **64** (1958) 571.
3. N. Bohr, *Library of Living Philosophers: A. Einstein* (1949) p. 201.

CHAPTER XV

IS QUANTUM MECHANICS A COMPLETE THEORY?

This problem was raised by A. Einstein, who gave an example that he considered to imply a negative answer. This example, known as the Einstein-Rosen-Podolski paradox[1], at one time engaged the attention of all physicists concerned with the basic problems of quantum theory. Nowadays it appears more difficult to formulate the "paradox" than to explain it, but we shall nevertheless examine it.

Suppose that the wave function of a system consisting of particles 1 and 2 has the form

$$\psi(x_1 - x_2) = \int e^{ip(x_1 - x_2 + a)}\, dp, \tag{1}$$

in which x_1 and x_2 are the coordinates of the particles; for these, p (the momentum) assumes the values $p_1 = p$ and $p_2 = -p$. This function may be put in the x representation as

$$\psi(x_1 - x_2) = \int \delta(x_1 - x_1')\, \delta(x_1' - x_2 + a)\, dx_1'. \tag{2}$$

We assume that we have measured the momentum of the first particle and found that $p_1 = p'$; then (1) implies that $p_2 = -p'$. Hence, although we have performed no action on the second particle, the packet of (1) has been contracted to the function

$$\exp\left[ip'(x_1 - x_2 - a)\right],$$

and the coordinate x_2 of particle 2 has become completely undefined. The same applies if the coordinate x_1 is measured: Let us suppose that we find $x_1 = x_1'$. Then (2) shows that $x_2 = a + x_1'$, i.e. the packet of (2) has contracted to $\delta(x_1' - x_2 + a)$. Here again we find that the measurement on the first particle has altered the state of the second, although by measuring coordinate x_1' we may (at least in principle) avoid any influence from the instrument on the second particle. Hence measurements on one particle alter the state of another, and that in such a manner as to alter the initial wave function $\psi(x_1 - x_2)$ in one of the following ways:

$$\psi(x_1 - x_2) \begin{cases} e^{ip'(x_1 + x_2 + a)} \\ \delta(x_1' - x_2 + a). \end{cases} \tag{3}$$

Thus we have established that: a) the state of particle 2 is altered no matter whether the measuring instrument has any influence on this particle, and b) the new states of particle 2 are mutually exclusive, since they are either states with defined momentum $p_2 = -p'$ or states with defined coordinate $x_2 = a + x_1'$. Einstein, Rosen, and Podolski based their approach on the then common view that a particle 'in reality' may have a coordinate and a momentum simultaneously, but that the disturbance due to the measuring instrument (unmeasurable disturbance in Bohr's terms) does not allow one to establish the simultaneous values of x and p. Therefore it was assumed that the concept of coexistence of coordinate x and momentum p must be eliminated from the theory, as not corresponding to the capacities of the observer.

The Einstein-Rosen-Podolski example has shown that the pair of canonically conjugate variables x and p cannot be measured together, in spite of the absence of any disturbance due to the instrument. The impossibility arises on account of the change in the state of particle 2, which arises without any action on it from the instrument. This lack of action means that there is no reason why x and p should not be measured simultaneously if this pair (x, p) occurs in a physical reality that quantum mechanics is, unfortunately, unable to describe in its language. Consequently, this would mean that quantum mechanics is not complete, which was the conclusion reached by the authors of the paradox.

This paradox has been discussed by N. Bohr[2] and L. Mandel'shtam[3] on the basis of different arguments, but ones that are not mutually in conflict.

Bohr based his reply to Einstein on the principle of complementarity; he stressed its significance as a new physical principle that rules out states of a particle having simultaneously defined canonically conjugate variables, e.g. x and p. From this viewpoint, no measurement can lead to a state with simultaneously defined values of x and p, no matter whether the means of measurement is direct or indirect (in the paradox, the variables x and p for the second particle are measured indirectly via the measurement on the first particle), and this is so no matter whether the instrument produces a disturbance or not. Such a measurement would contradict the principle of complementarity.

There is nothing wrong with Bohr's explanation, but Mandel'shtam's interpretation of the paradox appears to reveal its essence more fully.

Mandel'shtam's interpretation is that the change in the state of the second particle "without disturbance from the instrument" is actually due to the correlation of the states of particles 1 and 2 in the initial ensemble; the established fact $p_1 = p'$ implies that $p_2 = -p'$ as a direct consequence of the correlation. Thus Mandel'shtam employs a statistical interpretation of quantum mechanics to explain the paradox. The significance of the correlation may be elucidated from a more general example of two particles. Let the state of particles 1 and 2 be characterized by their belonging to a quantum ensemble described at $t=0$ by the wave function

$$\psi(x_1, x_2, 0) = \psi_a(x_1)\,\psi_b(x_2), \tag{4}$$

in which x_1 and x_2 are the coordinates of the particles, a and b being the values of certain other dynamic variables whose specification serves to define the initial ensemble.

The interaction between the particles gives rise to a new state for $t>0$, which is described by the wave function $\psi(x_1, x_2, t)$. We expand this as a spectrum with respect to the proper functions of operators A and B that represent the dynamic variables a and b:

$$\psi(x_1, x_2, t) = \int C(a', b')\,\psi_{a'}(x_1)\,\psi_{b'}(x_2)\,\mathrm{d}a'\,\mathrm{d}b'. \tag{5}$$

This spectral expansion shows that a measurement $a=a'$ allows us to say nothing about the value of variable b and hence about the state of particle 2 after the measurement on particle 1, which gave the result $a=a'$. The situation is altered if the system consisting of particles 1 and 2 is subject to some special laws; e.g. a and b may be subject to a law of conservation such that

$$\frac{d}{dt}(\mathscr{A} + \mathscr{B}) = [\mathscr{H}, (\mathscr{A} + \mathscr{B})] = 0, \tag{6}$$

in which $\mathscr{H}$ is the hamiltonian of system $1+2$. Here the coefficients

106

$C(a', b', t)$ in (5) take the form

$$C(a', b', t) = d(a', b', t)\, \delta(a' + b' - a - b), \qquad (7)$$

which indicates that the sum $a' + b'$ must be conserved, and that in this case (5) is replaced by

$$\psi(x_1, x_2, t) = \int d(a', a + b - a', t)\, \psi_{a'}(x_1)\, \psi_{a+b-a'}(x_2)\, da', \qquad (8)$$

so that a measurement giving $a = a'$ automatically implies $b' = a + b - a'$. The correlation of the states of particles 1 and 2 in this example is a consequence of the laws of interaction of the particles, i.e. the law of conservation for the quantity described by the operator $(A + B)$.

Any measuring device that splits up the initial state into a spectrum in terms of the states

$$\psi_{a'}(x_1)\, \psi_{b'}(x_2),$$

automatically gives a spectrum in terms of the states

$$\psi_{a'}(x_1)\, \psi_{a+b-a'}(x_2),$$

since there is no state with $b' \neq a + b - a'$ in the initial ensemble.

There is a strict relation between the values of a and b on account of the dynamics of the interaction. In the paradox we have $a = p_1$ and $b = p_2$, with $p_1 + p_2 = 0$ at $t = 0$. The form of $\psi(x_1, x_2)$ in Einstein's example (1) implies that momentum is conserved in the interaction of particles 1 and 2, which provides the correlation between the states of particles 1 and 2.

The paradox thus arose from neglect of the statistical character of quantum mechanics and also of the statistical correlation between the states of particles 1 and 2 consequent on the dynamics of the interaction.

Imagine a society whose laws of inheritance happen to be such that fair-haired people have only blue eyes while dark-haired people have only brown eyes. There is then no implication that the subject tends to be coloured by the mere act of selection in the observation that by selecting an individual with blue eyes we automatically find light hair in that individual; the observed correlation of hair and eye colours is a result of the laws of inheritance in that population.

Returning to the Einstein-Rosen-Podolski paradox, we see that there is no mysticism in the apparent influence of the instrument on the state of the particle, which in fact is not affected; the apparent "influence" is produced within the system by its own internal laws and is not a consequence of "interference" from the instrument.

Here we may recall that we have repeatedly stressed: a microparticle always interferes with the state of the instrument, whereas the instrument sometimes does, and sometimes does not, interfere with the state of the particle. Hence we must accept as unsatisfactory the idea that quantum mechanics is a direct consequence of the capabilities of macroscopic instruments; quantum mechanics reflects objective laws that govern the microworld but speaks of them in the language of the macroscopic world. In particular, it is impossible to measure x and p simultaneously for a microparticle as a consequence of the basic laws of quantum mechanics, and any instrument is bound by these: the laws state that the pair x and p does not relate to "physical reality" and so cannot be observed under any conditions.

This means that we must reject the assumption in the paradox that the pair x and p actually exists and that only the incompleteness of quantum mechanics prevents us from bringing the pair together.

REFERENCES

1. A. Einstein, B. Podolski, and N. Rosen, *Uspekhi fiz. nauk* **16** (1936) 436.
2. N. Bohr, A Reply to Einstein, *ibid.* **16** (1936) 446.
3. L. I. Mandel'shtam, Lectures on Quantum Mechanics. *Collected Works* [in Russian], vol. 5, Izd. AN SSSR 1950.

CHAPTER XVI

LATENT PARAMETERS

The abstract form of classical mechanics allows one to predict unambiguously the future of a system if the initial data are known. There is no need to stress how illusory is faith in such omnipotence of classical mechanics. However, this is precisely the basis on which classical statistical mechanics is often considered as a theory of the second kind, to which it is necessary to resort when a mechanical system becomes too complex for the equations of mechanics to be applied directly.

Physicists and philosophers who take this view of statistical mechanics usually consider that, in principle, the strictly deterministic equations of classical mechanics could be applied; thus statistical mechanics is viewed simply as a device for avoiding an excessively complex computational problem.

This naive viewpoint is difficult to reconcile with the fact that dynamics and statistics are inseparable in quantum mechanics, and the latter does not allow the most powerful of mathematicians (or computers) to abandon the statistical description of the microworld even in principle. There is a tendency to believe that quantum mechanics is merely a statistical method for the description of microsystems that "in fact" are subject to dynamical laws if they are described in terms of variables unknown in quantum mechanics, namely the latent parameters λ. Specification of these parameters would then eliminate statistics from quantum theory. It would be more accurate to say that the statistics of a quantum ensemble would arise as a result of inexact specification of these remarkable parameters.

This problem of the latent parameters λ has been discussed repeatedly; von Neumann appears to be the first to have directed attention to the problem. His outstanding monograph on the mathematical principles of quantum mechanics has had a great influence on the development of the idea of quantum ensembles in the USSR; von Neumann endeavoured to show that latent parameters are incompatible with the basic principles of quantum mechanics.[1] All the same, it appears that complete clarity was not attained, and that the problem demands further analysis.

The problem is one extremely difficult to discuss in any very general form. In one of his papers the present author has pointed out that it would be groundless to assume that thermodynamics may contain within itself the basis for doubting the absolute significance of the second law of thermodynamics, for that law lies at the very basis of the theory. However, progress in the kinetic theory showed that there are cases where the second law is explicitly violated without other conflicts with thermodynamics, as for example, in Brownian motion. The contradiction in this case was removed by adopting a wider view on the second law, taken in the kinetic theory of matter, namely the conception of the entropy S as a quantity defining the probability W of a state: $S = k \ln W$, in which k is Boltzmann's constant.

Another possibility to examine is the introduction of latent parameters such as to give meaning to a proportion of the form

$$\frac{x}{\text{quantum mechanics}} = \frac{\text{kinetic theory of matter}}{\text{thermodynamics}} \tag{1}$$

in which x is some unknown (more complete) theory.

It cannot be denied that the symbolic equation of (1), or some similar one, might be soluble, at least in this extremely general and purely methodological formulation of the problem. In what follows we shall attempt to select some narrower but more closely defined ways of discussing these latent parameters, which we may divide into two possible classes:

A) The parameters λ are variables observable in principle but not included among the purely dynamic variables considered in quantum mechanics;

B) The parameters λ are unobservable, even in principle.

These two classes must be discussed entirely separately.

A. OBSERVABLE LATENT PARAMETERS

Assume that the variables characterizing the macroscopic setting M and the dynamic variables of a microparticle (L, $\mathscr{P}$, Q, etc.) are accompanied by additional variables λ (latent parameters), the latter allowing one to complete the definition of the state of a quantum system in such a way as to reduce the random spread found with the usual dynamic and other variables of the quantum ensemble $M + \mu$.

Von Neumann distinguished two types of statistical ensemble: inhomogeneous and homogeneous. In the first type the mean value $\bar{A}$ of a positive random quantity A may be expressed as a sum:

$$\bar{A} = \alpha_1 \bar{A}_1 + \alpha_2 \bar{A}_2, \tag{2}$$

in which

$$\alpha_1 + \alpha_2 = 1, \quad \alpha_1, \alpha_2 > 0$$

and $\bar{A}_1$ and $\bar{A}_2$ are means taken over subensembles via some particular mode of selection of systems present in the ensemble, i.e. selection by reference to the features that determine the probabilities α_1 and α_2.

For a homogeneous ensemble subject to any mode of selection we have

$$\bar{A} = \bar{A}_1 = \bar{A}_2 = \cdots \tag{3}$$

A mixed quantum ensemble is clearly inhomogeneous. A very simple example of a mixed ensemble is the one formed by particles derived from two incoherent sources (e.g. the α-rays from two radioactive samples). Any mean relating to these particles is to be represented as in (2), α_1 and α_2 having the meaning of the probabilities that the particles come from the two sources.

This expansion is impossible, by virtue of the definition of a pure ensemble, if the ensemble is described by a single wave function Ψ_M and hence is a pure ensemble. The definition of the latter implies that it consists of an infinite set of identical macroscopic settings M, which produce the conditions of existence for the microparticles μ. Set M, of number $N \to \infty$, may be imagined as divided into two subsets

$$N_1 + N_2 = N, \quad \frac{N_1}{N_2} = \frac{\alpha_1}{\alpha_2}$$

each subset N_1 or N_2 having identical M and μ. By identical is meant here that the macroscopic parameters defining M are the same, while macroscopically measurable parameters of the μ (such as mass m, charge e, spin σ, etc.) are also identical. Then the Ψ_M that defines the quantum ensemble for subset N_1 or N_2 is precisely the same, by virtue of the definition of the wave function itself, and so for any measurable quantity A we have $\bar{A}_1 = \bar{A}_2$; hence the decomposition as in (2) is impossible.

This conclusion cannot conform with the possibility of introducing latent parameters λ, since the latter are essentially such as would, if measured, allow us to "define more closely" the quantum variables, e.g. they would allow us to alter the variance of some dynamic variable L. For instance, suppose that in the range $J_1(\lambda)$ of the latent parameters the mean $\bar{L}$ takes the value L_1, with variance $\overline{\Delta L_1^2}$, while in region $J_2(\lambda)$ the corresponding values are L_2 and $\overline{\Delta L_2^2}$. Then $\bar{L}$ and L^2 for the entire ensemble are

$$\bar{L} = \alpha_1 L_1 + \alpha_2 L_2, \tag{4}$$

$$\overline{\Delta L^2} = \alpha_1 \cdot \overline{\Delta L_1^2} + \overline{\alpha_2 L_2^2} + \alpha_1 (L_1 - \bar{L})^2 + \alpha_2 (L_2 - \bar{L})^2, \tag{4'}$$

in which α_1 and α_2 are the relative weights of the regions $J_1(\lambda)$ and $J_2(\lambda)$, with

$$G_2(\lambda), \quad \alpha_1 + \alpha_2 = 1, \quad \alpha_1, \alpha_2 \geqslant 0.$$

In other words, a pure ensemble is inhomogeneous (in the parameters λ), which is contrary to the definition.

If now we seek to define more closely these λ by dividing the entire region of possible values of λ into smaller parts

$$G_1(\lambda), G_2(\lambda), ..., G_s(\lambda), ..., G_N(\lambda)$$

with weights

$$\alpha_1, \alpha_2, ..., \alpha_s, ..., \alpha_N, \sum_1^N \alpha_s = 1$$

then we find for the limit of infinitely small regions $J_s(\lambda)$ that (4') is replaced by

$$\overline{\Delta L^2} = \sum_\lambda \alpha_\lambda (L_\lambda - \bar{L})^2, \tag{5}$$

in which

$$\alpha_\lambda = \frac{G_s(\lambda)}{\sum_{s=1}^N G_s(\lambda)}$$

and L_λ is the mean of L in $J_s(\lambda)$, this mean equalling the exact value of L in region $J_s(\lambda)$ for $N \to \infty$.

Formula (5) shows that the entire random spread in L is now due to the uncertainty in the λ; the quantum dynamic variables should take completely definite values if these parameters are determined accurately, i.e. if $J_s(\lambda)$ is a narrow region.

This leads to further contradictions; there are no microsystems that can be described in terms of space-time (Q-type) variables alone or in terms of momentum-energy (P-type) ones alone. The principle of complementarity indicates that the operators representing these quantities are not commutative:

$$\mathscr{P}\mathscr{Q} - \mathscr{Q}\mathscr{P} = i\hbar, \tag{6}$$

so $\overline{\Delta P^2} \neq 0$ if $\overline{\Delta Q^2} = 0$ in a quantum ensemble defined by the macro-setting M; similarly, $\overline{\Delta Q^2} \neq 0$ if $\overline{\Delta P^2} = 0$. Hence if some dynamic variable L of a quantum ensemble is determined exactly $(\overline{\Delta L^2} = 0)$, we have either $\overline{\Delta P^2} > 0$, or $\overline{\Delta Q^2} > 0$, or both simultaneously. Hence we would arrive at a conflict with the principle of complementarity if we were to find a range $J(\lambda)$ of the latent parameters that would give a subensemble with $\overline{\Delta Q^2} = 0$ and $\overline{\Delta P^2} = 0$.

The following argument[2] also confirms the difficulty of introducing these latent parameters λ. Let the relevant dynamic variable be the electron spin $\mathbf{S} = (S_x, S_y, S_z)$. We have

$$[S_x, S_y] = i\hbar S_z. \tag{7}$$

The two possible values for the variables S_x, S_y and S_z are $\pm\hbar/2$. We assume that it has been established that the projection of the spin on the OZ axis is $S_z = +\hbar/2$. Let this value of the quantum variable be a consequence of the fact that the λ lie in the range $J_z^+(\lambda)$.

We now alter the total system $\mathfrak{M} = \mathscr{M} + (A+D)_z$, replacing $(A+D)_z$ by $(A+D)_x$, thereby sorting the particles μ in accordance with the variable S_x to get values of the latter equal to $+\hbar/2$ and $-\hbar/2$. Let $J_x^+(\lambda)$ be the range in λ giving $S_x = +\hbar/2$. Then $J_z^+(\lambda)$ splits up as

$$G_z^+(\lambda) = G_x^+(\lambda) + G_x^-(\lambda). \tag{8}$$

Suppose we have found that $S_x = +\hbar/2$. This means that the λ lie in the range $J_x^+(\lambda)$. We now again measure the projection of $\mathbf{S}$ on the OZ axis, for which purpose we replace $(A+D)_x$ by the starting $(A+D)_z$. We then obtain values of $+\hbar/2$ and $-\hbar/2$ for S_z, so that

$$G_x^+(\lambda) = G_z^+(\lambda) + G_z^-(\lambda). \tag{9}$$

Comparison of (8) with (9) gives us the contradiction

$$G_x^+ (\lambda) + G_z^- (\lambda) = 0.\tag{10}$$

We may escape formally from this if we assume that the λ relate to the measuring instrument $A+D$, not to $\mathcal{M}+\mu$; on replacing $(A+D)_x$ by $(A+D)_z$ or vice versa, the λ change their physical significance. The latent parameters when S_z is measured are not those when S_x is measured, so that λ on the right and left in (8) are physically different. In other words, the λ vary with the measuring devices $A+D$, here $(A+D)_x$ and $(A+D)_z$. The above possibility cannot be rejected *a priori*, but this view of the λ does not refer them to the class of quantities that supplement the dynamic variables of quantum mechanics, for they obviously belong to the class of quantities considered in quantum mechanics, e.g. the variables that characterize the measuring instrument, whose essence is such as to produce a unique relation between the measured quantity L and the range $J(\lambda)$ of the λ characteristic of the device. That is, if $\lambda \in J_{L'}(\lambda)$, then the dynamic variable equals L'.

Thus we see that it is impossible to introduce latent parameters λ that are in principle observable into section $M+\mu$ of $\mathfrak{M}$ without causing a conflict with the principle of complementarity.

B. UNOBSERVABLE LATENT PARAMETERS

We now assume that the λ are not observable even in principle. The term "in principle" recalls the notorious "the basis of the principial observability" that some have considered as the philosophical basis of quantum mechanics[3], and it must therefore be stressed that the present "in principle" implies a definite theory and its principles, which allow some things and forbid others. In discussing "essentially unobservable latent parameters" we are also obliged to have in mind the principles of some definite theory, not a vast ocean of possibilities limited only by the power of imagination. As our theoretical basis of discussion we take quantum mechanics and assume that the term "essentially unobservable quantity" defines a quantity whose observability is forbidden by the principles of quantum mechanics. This conception of the λ does not conflict with the principles of quantum mechanics.

Consider now the latent parameters λ of a pure quantum ensemble; if these are unobservable, we cannot point to any method that would allow us to isolate from the pure ensemble any subensemble whose means $\bar{L}$ and $\bar{L^2}$ differ from those characteristic of the ensemble as a whole. That is, a decomposition as in (4) or (4') is not feasible, so there is no conflict with the definition of pure ensemble, nor any with uncertainty relationships of the form $\overline{\Delta P^2} . \overline{\Delta Q^2} \geqslant a^2$, given these assumptions about the λ. Moreover, we shall show that such unobservable parameters λ occur in current quantum mechanics. Henceforth we shall follow R. Feynman's lines of thought[4]; Feynman has shown that the wave function describing a pure ensemble may be represented as an integral over "ideal" (unobservable) particle trajectories $x = x(t)$, $dx/dt = \dot{x}(t)$, in which the $x(t)$ are the coordinates of the particle at times t and $\dot{x}(t)$ is the velocity.

Feynman's approach may be presented by recalling that a quantum ensemble may be regarded as a series of unitary transformations. Let the state of the microparticle be represented by the dynamic variable a, which has a continuous spectrum. Let further $\psi(a_0, t_0)$ be the wave function describing the state of the ensemble at time $t = t_0$, in which a_0 is the value of a at $t = t_0$ (from the definition of the wave function). Then the wave function $\psi(a, t)$ for $t > 0$ may be put in the form

$$\psi(a, t) = \int U(a, t; a_0, t_0) \, \psi(a_0, t_0) \, da_0 , \tag{11}$$

in which

$$U(a, t; a_0, t_0)$$

is the unitary matrix for the canonical transformation from the variables a_0 to the variables a, which satisfies Schrödinger's equation:

$$i\hbar \frac{\partial U}{\partial t} = \mathcal{H} U , \tag{12}$$

in which $\mathcal{H}$ is the hamiltonian of the system. If for U we take initial conditions such that

$$U = \delta(a - a_0) \quad \text{for} \quad t = t_0 , \tag{13}$$

$$U = 0 \quad \text{for} \quad t < t_0 , \tag{13'}$$

then this U coincides with the Green's function G for (12) considered earlier.

We take in (11) some time t_1 such that $t_0 < t_1 < t$ in place of the arbitrary t; then (11) transforms $\psi(a_0, t_0)$ into $\psi(a_1, t_1)$. We can now take $\psi(a_1, t_1)$ as the initial function and use $U(a, t; a_1, t_1)$ to obtain once more the function $\psi(a, t)$ for $t > t_1$; we obviously get

$$\psi(a, t) = \int da_1 \int da_0 \, U(a, t; a_1, t_1) \, U(a_1, t_1; a_0, t_0) \, \psi_0(a_0, t_0). \quad (14)$$

Repeating this procedure for intermediate points in time

$$t_0 < t_1 < t_2 < \cdots < t_N = t$$

and for the values of the dynamic variable

$$a_1, a_2, \ldots, a_N = a_t = a,$$

corresponding to these instants, we can represent the transformation matrix $U(a, t; a_0, t_0)$ in the form

$$U(a, t; a_0, t_0) = \int \cdots \int da_1 \, da_2 \ldots da_{N-1} \, U(a_2 t; a_{N-1}, t_{N-1})$$
$$\times U(a_{N-1}, t_{N-1}; a_{N-2}, t_{N-2}) \ldots U(a_2, t_2; a_1, t_1)$$
$$\times U(a_1, t_1; a_0, t_0), \quad (15)$$

i.e., as a series of "transitions"

$$a_0 \to a_1 \to a_2 \to \cdots \to a_{N-1} \to a_N$$

for possible intermediate values of the variables $a_1, a_2, \ldots, a_{N-1}$. This formula is analogous to the classical formula[5] for a Markoff chain:

$$P(a, t/a_0, t_0) = \int \cdots \int da_1 \, da_2 \ldots da_{N-1} \, \mathscr{P}(a, t; a_{N-1}, t_{N-1})$$
$$\times P(a_{N-1}, t_{N-1}; a_{N-2}, t_{N-2}) \ldots P(a_2 t_2; a_1, t_1)$$
$$\times P(a_1, t_1; a_1, t_0), \quad (16)$$

in which

$$P(a_k, t_k, a_{k-1}, t_{k-1})$$

is the probability of a transition from state a_{k-1} to state a_k in the period $t_k - t_{k-1}$. Formula (16) is the probability of transition from state a_0 at time t_0 to state a at time $t(t>t_0)$ as a result of all possiblet ransitions via intermediate values of a.

The quantum formula (15) has the probability P replaced by the probability amplitude U which is related to P by

$$P(a_k, t_k; a_{k-1}, t_{k-1}) = \left| U(a_k, t_k; a_{k-1}, t_{k-1}) \right|^2. \tag{17}$$

This distinction means that the externally similar formulae (15) and (16) are fundamentally different.

In relation to (14) and (15), we assume that a is the cartesian coordinate x of the microparticle; we put $t_{k+1}-t_k=\Delta t$ and assume that the transformation function U for small Δt may be put as

$$U(x_{k+1}, t_{k+1}; x_k, t_k) = \frac{1}{A} \exp\left[\frac{i}{\hbar} S(x_{k+1}, x_k, \Delta t)\right], \tag{18}$$

in which A is a normalization factor (independent of x_k and t_k) and

$$S(x_{k-1}, x_k, \Delta t)$$

is the phase of U. This phase is exactly the classical action function in the case of free motion, so we may assume that it remains correct also in the presence of an external field described by a potential $V(x)$. We suppose that this is true; then

$$S(x_{k+1}, x_k, \Delta t) = \left\{\frac{m}{2}\left(\frac{x_{k+1}-x_k}{\Delta t}\right)^2 - V(x_{k+1})\right\} \Delta t, \tag{19}$$

in which m is the mass of the microparticle. In the limit of infinite subdivision of the interval (t_0, t) we get

$$S(\dot{x}, x, dt) = \{\tfrac{1}{2}m\dot{x}^2 - V(x)\}\, dt. \tag{20}$$

This is the element of classical action in time dt;

$$\dot{x}(t) = \lim_{\Delta t \to 0} \frac{x_{k+1}-x_k}{\Delta t}$$

is the velocity of the particle at time t, and

$$\mathscr{L}(\dot{x}, x) = \tfrac{1}{2}m\dot{x}^2 - V(x)$$

is the classical Lagrange function. We shall show that the $\psi(x, t)$ given

117

by (11) subject to (18) and (19) satisfies the Schrödinger equation

$$i\hbar \frac{\partial \psi(x, t)}{\partial t} = \left\{ -\frac{\hbar^2}{2m} \frac{\partial^2}{dx^2} + V(x) \right\} \psi(x, t). \qquad (21)$$

The proof* may be given by considering the transformation from $\psi(x, t)$ to $\psi(x, t + \Delta t)$:

$$\psi(x, t + \Delta t) = \frac{1}{A} \int \exp\left[\frac{i}{\hbar} \left\{ \frac{m}{2} \left[\frac{x_{N+1} - x_N}{\Delta t} \right]^2 - V(x_{N+1}) \right\} \Delta t \right]$$
$$\psi(x_N, t)\, dx_N. \qquad (22)$$

We put

$$x_{N+1} - x_N = \xi, \quad x_{n+1} = x,$$

so (22) is replaced by

$$\psi(x, t + \Delta t) = \frac{1}{A} \int \exp\left[\frac{i}{\hbar} \frac{m}{2} \frac{\xi^2}{\Delta t} - \frac{i}{\hbar} V(x)\, \Delta t \right] d\xi\; \psi(x - \xi, t). \qquad (22')$$

We expand the

$$\psi(x - \xi, t) = \psi(x, t) - \frac{\partial \psi(x, t)}{\partial x} \xi + \tfrac{1}{2} \frac{\partial^2 \psi(x, t)}{\partial x^2} \xi^2 + \cdots$$

appearing here and perform a simple integration with respect to ξ to get

$$\psi(x, t) + \frac{\partial \psi(x, t)}{\partial t} \Delta t + \cdots$$
$$= \exp\left[-\frac{i}{\hbar} V(x)\, \Delta t \right] \left(\frac{2\pi i\hbar\, \Delta t}{m} \right)^{\frac{1}{2}} \frac{1}{A}$$
$$\left\{ \psi(x, t) + \frac{i\hbar}{m} \frac{\partial^2 \psi(x, t)}{\partial x^2} \Delta t + 0(\Delta t^2) \right\} \qquad (22'')$$

We take our normalization factor A as equal to

$$\left(\frac{2\pi i\hbar\, \Delta t}{m} \right)^{\frac{1}{2}}$$

* This proof is due to Feynman[4]; Dirac[6] appears to have been the first to point out the relation of the phase S to the classical action $\mathscr{L}$.

and perform the series expansion

$$e^{-i/\hbar V(x)\Delta t} = 1 - i/\hbar \cdot V(x)\,\Delta t + \cdots,$$

to demonstrate that $\psi(x, t)$ for $t \to 0$ actually does satisfy (21). This allows us to put (15) as

$$U(x, t; x_0, t_0) = \int \cdots \int \exp\left[\frac{i}{\hbar} \sum_{k=1}^{N} S(x_k, x_{k-1}, \Delta t)\right]$$
$$\frac{dx_N}{A} \frac{dx_{N-1}}{A} \cdots \frac{dx_k}{A} \cdots \frac{dx_1}{A}, \qquad (23)$$

in which

$$S(x_k, x_{k-1}, \Delta t)$$

is the classical action in time Δt given by (19).

If we replace $1/\hbar$ by

$$\frac{1 + i\delta}{\hbar}, \quad \delta > 0,$$

the integral of (23) has the meaning of a functional integral with respect to a Wiener metric also for

$$N \to \infty, \quad \Delta t = \frac{t - t_0}{N} \to 0,$$

The integral of (23) for $N \to \infty$ is the limit of this functional integral for $\delta \to 0$ (see ref. 7); as $\Delta t \to 0$ the sum

$$\sum_{k=1}^{N} S(x_k, x_{k-1}, \Delta t)$$

is shown by (10) to go over to the integral of the classical action in the period from t_0 to t:

$$\int_{t_0}^{t} \mathscr{L}[\dot{x}(\tau), x(\tau)]\,d\tau = S(x, t, x_0, t_0),$$

the volume

$$\frac{dx_N}{A} \frac{dx_{N-1}}{A} \cdots \frac{dx_1}{A}$$

for $N \to \infty$ being put symbolically as $dW\{x(\tau)\}$. Then (23) is replaced by

$$U(x, t; x_0, t_0) = \int dW\{x(\tau)\} \exp\left[\frac{i}{\hbar} \int_{t_0}^{t} \mathscr{L}(\dot{x}(\tau), x(\tau))\, d\tau\right]. \tag{24}$$

In other words, the transformation function (probability amplitude)

$$U(x, t; x_0, t_0)$$

may be represented as a functional integral over all trajectories leading from point (x_0, t_0) to point (x, t). Each trajectory

$$x(\tau), \dot{x}(\tau), t_0 \leqslant \tau \leqslant t$$

gives a contribution to

$$U(x, t; x_0, t_0)$$

equal to

$$dU = dW\{x(\tau)\} \exp\left[\frac{i}{\hbar} \int_{t_0}^{t} \mathscr{L}(\dot{x}, x)\, d\tau\right]. \tag{25}$$

The constancy of A means that the amplitude $dW\{x(\tau)\}$ for each trajectory is the same, whereas the phases

$$\frac{i}{\hbar} \int_{t_0}^{t} \mathscr{L}(\dot{x}, x)\, d\tau$$

are different. The factors

$$\exp\left[\frac{i}{\hbar} S(x, t, t_0)\right],$$

$$S(x, t, t_0) = \int_{t_0}^{t} \mathscr{L}(\dot{x}, x)\, d\tau,$$

are oscillatory, so the largest contribution to (24) clearly arises from terms of the form of (25) for which $S(x, t, t_0)$ is minimal, i.e. from those trajectories for which

$$\delta \int_{t_0}^{t} \mathscr{L}(\dot{x}, x)\, \mathrm{d}\tau = 0, \tag{26}$$

where δ denotes a variation of trajectory $x(\tau)$. However, condition (26) is also the condition that defines the classical trajectory of the particle, so we may interpret as follows Feynman's striking result represented by formulae (24), (25), and (26): a microparticle prefers to move along the classical trajectory $x(\tau)$, which satisfies the principle of least action expressed by (26), but it does not avoid the other possible trajectories leading from (x_0, t_0) to (x, t) (Figure 11); (24) shows that all of these

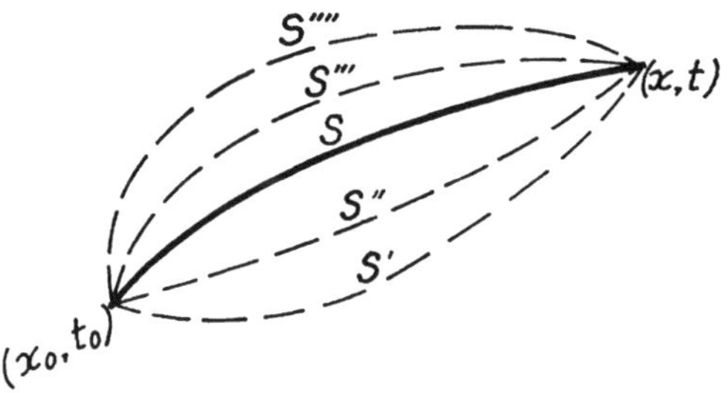

Fig. 11.   Classical trajectory S of a particle corresponding to a minimum in the action function along S; paths S', S'', etc. are ideal quantum ones, which also play a part in the motion of a microparticle.

possible trajectories make a definite contribution to the motion of the quantum ensemble. However, the "ideal" trajectories $\dot{x}(t)$ and velocities $x(t) = p(t)/m$ (in which p is the momentum of the particle) appearing in (27) and (14) are unobservable, because the basic principles of quantum mechanics indicate that there is no quantum ensemble in which the variables x and $\dot{x} = p/m$ would be defined simultaneously.

Thus the $x(t)$ and $\dot{x}(t)$ of (24) provide an example of latent parameters, and unobservable ones at that; however, we have seen that there is no conflict between the principles of quantum mechanics and the existence of these latent parameters $\lambda = x(t)$, $\dot{x}(t)$ that determine the "ideal" path of a microparticle.

121

REFERENCES

1. J. von Neumann, *The Mathematical Basis of Quantum Mechanics* [Russian translation], Nauka, 1964.
2. D. Blokhintsev, *Uspekhi fiz. nauk* **25** (1951) 195.
3. D. Blokhintsev, *ibid.*
4. R. Feynman, *Causality in Quantum Mechanics* [Russian translation], IL, 1955; Rev. Mod. Phys. **20** (1948) 367.
5. E. B. Dynkin, *Markoff Processes* [in Russian], Fizmatgiz, 1963.
6. P. A. M. Dirac, *The Principles of Quantum Mechanics*, section 32, 3rd ed., 1947.
7. I. M. Gel'fand and A. M. Yaglom, *Uspekhi matem. nauk* **11** (1956) 76.

CHAPTER XVII

CAN A PARTICLE HAVE AN INDIVIDUAL HISTORY?

Some people tend to be disturbed by the failure of current quantum mechanics to describe the fate of a single electron. This concern for the fate of a single microparticle, a single individual in the microworld, seems rather hypocritical for even in the macrocosm the fate of a given object cannot be predicted over a prolonged period.

Perhaps this desire for prediction derives from the natural longing of man to know his future, which is sometimes so acute that some people even resort to fortune tellers. However, it is not difficult to appreciate the psychological state of a thinking man who would know precisely the entire course of his life; would it not lose for him all interest, as does a film whose plot is known in all its details?

This comment has no direct application to the problems of quantum mechanics, but it may tend to assure those who continually repeat that the future cannot be predicted unambiguously, not even for an elementary particle.

Consider now the history of an individual particle in the microcosm. First we must define more closely what we mean by this history, and what the history of a particle might signify. It is clear that it is correct to view the history of a particle as a sequence of its states in time, so we must start with the concept of the state of a particle.

We know from the principles of quantum theory that we can describe the state of a particle via space-time variables (denoted by Q) or via momentum-energy ones (denoted by P); each set Q or P must form a complete set of dynamic variables, i.e. they must be represented by commutative operators

$$[\mathcal{Q}_s, \mathcal{Q}_r] = 0, \quad s, r = 1, 2, ..., f, \tag{1}$$

or, for the momentum set,

$$[\mathcal{P}_s, \mathcal{P}_r] = 0, \quad s, r = 1, 2, ..., f, \tag{1'}$$

and these must be mutually independent, their number equalling f, the number of degrees of freedom of the microsystem. The state of a particle is thus specified by the complete set of Q-type dynamic variables: $Q_1, Q_2, ..., Q_f$ or alternatively $(P) = P_1, P_2, ..., P_f$.

Prediction of the history of a particle would mean prediction of the sequence $Q(t)$ or $P(t)$ for instants t:

$$t: t_1 < t_2 < \cdots < t_k < \cdots < t_N.$$

An interesting theorem due to Mandel'shtam[1] may be considered before we turn to an analysis of the problems arising here.

In the general uncertainty relation of (XI–3) we put $\mathscr{A} = \mathscr{L}$ and $\mathscr{B} = \mathscr{H}$, in which $\mathscr{H}$ is the operator for the energy E; we then have

$$\overline{\Delta L^2} \cdot \overline{\Delta E^2} \geqslant \frac{\hbar^2}{4} \, |\overline{[\mathscr{H}, \mathscr{L}]}|^2 , \tag{2}$$

in which $\overline{\Delta L^2}$ and $\overline{\Delta E^2}$ are the variances of L and E, while

$$[\mathscr{H}, \mathscr{L}] = \frac{1}{i\hbar} (\mathscr{H}\mathscr{L} - \mathscr{L}\mathscr{H}).$$

On the other hand,

$$\overline{[\mathscr{H}, \mathscr{L}]} = \frac{\mathrm{d}L}{\mathrm{d}t} . \tag{3}$$

Hence,

$$\overline{\Delta L^2} \cdot \overline{\Delta E^2} \geqslant \frac{\hbar^2}{4} \left| \frac{\overline{\mathrm{d}L}}{\mathrm{d}t} \right|^2 . \tag{4}$$

This formula implies that we have an "uncertainty relation" of the following form if the mean $\bar{L}$ of L in time Δt alters by an amount $\overline{\Delta L} \approx \overline{(\Delta L^2)}/^{\frac{1}{2}}$:

$$\Delta E \cdot \Delta t \geqslant \frac{\hbar}{2} , \tag{5}$$

in which

$$\Delta E = \sqrt{\overline{\Delta E^2}} .$$

Consider now the description of the history of the particle in terms of the $\mathscr{P}$ variables. Let one of these variables, say $\mathscr{P}_f$, be equal to $\mathscr{H}$ then $\Delta E = 0$, and (5) shows that we require an infinite time Δt for an appreciable change to occur in the mean value of any of the dynamic variables $(\varDelta t \geqslant \hbar/2\varDelta E)$.

Thus there is no history at all if the microparticle is described via the momentum-energy variables. Of course, this conclusion is not unexpected because the description of the state via these variables would correspond in classical mechanics to description of the motion by means of integrals of motion, which are essentially independent of time.

An entirely different situation arises if the history of a particle is described via the complete set of Q-type variables, which do not commutate with the hamiltonian $\mathscr{H}$, by virtue of the principle of complementarity; hence generally speaking the means of dQ/dt differ from zero. Then (4) implies that $\overline{\varDelta E^2} > 0$ and:

$$\overline{\varDelta E^2} \geqslant \frac{\hbar^2}{4}\left(\overline{\frac{dQ}{dt}}\right)^2 \cdot \frac{1}{\varDelta Q_2}, \tag{6}$$

so that states of the ensemble are not stationary if these Q variables are specified with finite error $(\overline{\varDelta Q^2} \neq 0)$; all the probabilities, means, variances, and so on for all of the quantities will vary with time. The ensemble has a history, and hence a microparticle has one too.

The prediction of the history of a microparticle will consist of a prediction of the probability of some definite trajectory for the microparticle:

$$(Q_0, t_0), (Q_1, t_1), ..., (Q_k, t_k), ..., (Q_N, t) \tag{7}$$

From (XVI–16) we have[2] that the probability of such a trajectory is

$$\begin{aligned}
dW(Q_0, t_0; Q_1, t_1; ...,& Q_k, t_k, ..., Q_N, t) = P(Q_N, t; Q_{N-1}, t_{N-1}) \\
&P(Q_{N-1}, t_{N-1}; Q_{N-2}, t_{N-2}), ..., P(Q_k, t_k; Q_{k-1}, t_{k-1}), ..., \\
&P(Q_1, t_1; Q_0, t_0)\, dQ_1, ..., dQ_{N-1}, \tag{8}
\end{aligned}$$

in which the individual factors are given by (XVI–17). For the very simple case of free motion we have from (XVI–19) that

$$P(Q_k, t_k; Q_{k-1}, t_{k-1})\, dQ_k = \frac{dQ_k}{A}\left|\exp\left[\frac{im}{\hbar}\frac{(x_k - x_{k-1})^2}{\varDelta t}\right]\right|^2 \tag{9}$$

in which

$$\varDelta t = t_k - t_{k-1}.$$

The product in (8) has the meaning of the probability that successive measurements of coordinate Q at instants

$$t_0 < t_1 < t_2, ..., < t_k < , ..., < t_N$$

will give respectively the values

$$Q_0 \pm \tfrac{1}{2}\,\mathrm{d}Q_0, \quad Q_1 \pm \tfrac{1}{2}\,\mathrm{d}Q_1, ..., Q_k \pm \tfrac{1}{2}\,\mathrm{d}Q_k, ..., Q_N \pm \tfrac{1}{2}\,\mathrm{d}Q_N$$

each such measurement being a macroscopic event. Hence the probability of (8) is the probability of a certain sequence of macroscopic events that serves to describe the history of an individual particle (Figure 12).

This chain of events could be extremely striking if each measurement of Q did not lead to a state of infinitely large energy E. In this case the

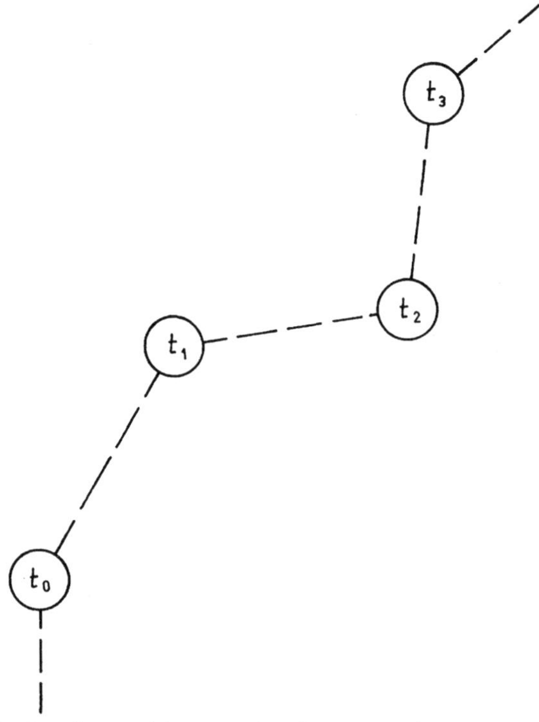

Fig. 12.   Path of a microparticle as a series of macroscopic events: each circle denotes a macroscopic phenomenon caused by the microparticle, e.g. a bubble in a bubble chamber.

operator for the energy of the microparticle is

$$\mathscr{H} = -\frac{\hbar^2}{2m}\frac{\mathrm{d}^2}{\mathrm{d}x^2}$$

while the wave function is

$$\psi(x, t) = \frac{1}{A}\exp\left[\frac{im}{\hbar\,\varDelta t}(x - x_0)^2\right], \tag{10}$$

in which x_0 is the coordinate derived from the previous measurement. The mean energy $\bar{E}$ of the microparticle in state (10) is given by the usual formulae of quantum mechanics as

$$\bar{E} = \frac{\int \psi^* \mathscr{H} \psi \,\mathrm{d}x}{\int \psi^* \psi \,\mathrm{d}x}. \tag{11}$$

These integrals are easily calculated if we replace the m of (10) by $m+i\delta$ with $\delta>0$, thereafter passing to the limit $\delta\to0$, which gives

$$\bar{E} = \frac{\hbar^2}{8m}\frac{m^2}{\delta}\cdot\frac{1}{\hbar\,\varDelta t}, \quad \delta\to+0, \tag{12}$$

so that $\bar{E}$ tends to $+\infty$ as $\delta\to+0$. Hence to produce the trajectory described by (8) we must have unlimited energy from some source: either in the measuring instrument $A+D$ or in the microparticle μ itself.

For this reason we shall now consider a more realistic scheme for the description of the trajectory of an individual particle, in which we do not assume an infinitely precise measurement of Q.

We assume that the microparticle interacts with infinitely heavy atoms whose disposition in space is known, thereby transferring them from the ground state $\psi_0(x)$ to an excited state $\psi_n(x)$, in which x is the coordinate of an electron in the atom, and $\psi(x)$ is its wave function. The interaction energy $W(x-Q)$ of our microparticle with an electron of the atom will be taken as represented by a δ-function:

$$W(x - Q) = g\cdot\delta(x - Q), \tag{13}$$

in which Q is the coordinate of the microparticle and g is the coupling constant. The atom receives an excitation energy $\varepsilon = E_n - E_0$, while the energy of the microparticle is $E = P^2/2M$, M being the mass of the particle, whose momentum before the collision is P and after the collision P'.

We assume that $E \gg \varepsilon$, so that the microparticle itself is the source of the high energy needed for its localization.

We suppose that the matrix element for the density of the electron in the atom,

$$\rho_n(x) = \psi_n^*(x)\,\psi_0(x) \tag{14}$$

vanishes for $|x| \gg a$, so that a behaves as the size of the atom. The Fourier transform of this is

$$\tilde{\rho}_n(q) = \int \rho_n(x)\,e^{iqx}\,\mathrm{d}^3x \tag{15}$$

and will vanish in the region $|q| \gg 1/a$.

For simplicity we shall also assume that only one matrix element $\rho_n(x)$ is of any importance, all the others being negligibly small.

The role of the analyser A is played by the excitation $E_0 \to E_n$ of the atom localized in space; this sorts the microparticles in accordance with their Q coordinates with an error of the order of a, the size of the atom. This excited atom may subsequently produce a latent-image state in a grain of a photographic emulsion, local evaporation in a bubble chamber, condensation in a Wilson cloud chamber, and so on (all macroscopic effects). These or other processes act as the detector D that destroys the interference between the various states of the microparticle.

The mode of operation of the detectors will not be discussed; we consider only the primary function of part A, which performs the analysis of the coordinates. The analyser A (here a heavy atom capable of being excited) is macroscopic because the mass of the atom has been assumed to be infinitely large, so that the position of its centre of gravity in space is completely defined.

The wave function of the system $\mu + A$ at the initial instant $t = 0$ is

$$\phi_0(x, Q) = \psi_0(x)\exp\!\left(i\,\frac{pQ}{\hbar}\right). \tag{16}$$

For $t > 0$ this function is put as

$$\phi\ x, Q, t) = \phi_0(x, Q)\,e^{i\omega_0 t} + \varphi(x, Q, t), \tag{17}$$

in which $\omega_0 = (E + E_0)/\hbar$; $\varphi(x, Q, t)$ will be taken as small and calculated via perturbation theory. The method used in Chapter XI gives us from (XI, 14) that

$$\varphi\ x, Q, t) = \frac{g\, e^{i\omega_0 t}}{i\hbar} \int \rho_n(x') \exp\left[\frac{i}{\hbar}(P - P', \bar{Q})\right] \delta(x' - Q')$$

$$\frac{1 - e^{-i\Omega t}}{\Omega}\, d^3 p' \cdot dx'\, dQ \cdot \psi_n(x) \exp\left[\frac{i}{\hbar}\, QP'\right], \tag{18}$$

$$\Omega = \frac{z}{t} = \frac{1}{\hbar}(E_P + E_0 - E_{P'} - E_n = E_P - E_{P'} - \varepsilon). \tag{19}$$

We assume that $E_P \gg \varepsilon$, so that we have

$$\Omega \cong \frac{1}{\hbar}(E_P - E_{P'}) = \Omega_{PP'}.$$

Consider now $\varphi(x, Q, t)$ for

$$0 < t \ll \frac{1}{\Omega_{PP'}}$$

i.e. immediately after a possible collision of the microparticle with the atom. We have from (18) that

$$\varphi\ x, Q, t) \cong t \cdot \frac{g \cdot e^{i\omega_0 t}}{\hbar}\, \psi_n(x) \int \tilde{\rho}_n(q')\, e\, \frac{iQP}{\hbar} + \frac{iQq}{\hbar}\, d^3 q, \tag{20}$$

in which $\mathbf{q} = \mathbf{P}' - \mathbf{P}$, or

$$\varphi(x, Q, t) = t \cdot \frac{g \cdot e^{i\omega_0 t}}{\hbar}\, \psi_n\ x)\, \rho_n(Q) \exp\left[\frac{iQP}{\hbar}\right] \cong \psi_n(x)\, \phi_n(Q). \tag{20'}$$

In other words, the state of the microparticle (which produces the excitation $E_0 \to E_n$) is represented by the following wave function for small t:

$$\phi_n(Q) = \rho_n(Q) \exp\left[\frac{i}{\hbar}\, QP\right]. \tag{21}$$

This function is a wave packet with a probability distribution for the coordinate of the microparticle:

$$W_n(Q) = |\rho_n(Q)|^2. \tag{22}$$

the initial average momentum being

$$\mathbf{P} = - i\hbar \int \phi_n^*(Q)\, \nabla_Q \phi_n(Q)\, dQ. \qquad (23)$$

This is so to within the momentum of the electron in the atom, which is of the order $\hbar/a \ll \mathscr{P}$.

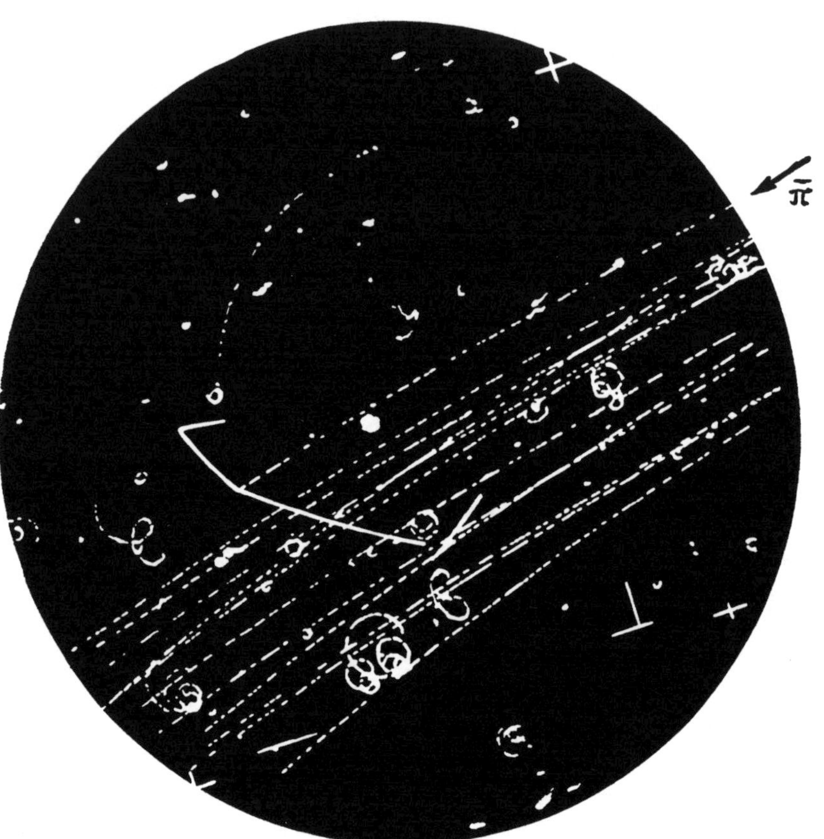

Fig. 13. Tracks of π^- mesons with an energy of 340 MeV in a hydrogen bubble chamber (recorded in the Nuclear Problems Laboratory Joint Nuclear Research Institute). The π^- meson denoted by the arrow interacts with a proton p via the reaction $\pi^- + p \rightarrow n + \pi^- + \pi^+$ (n is a neutron), the π^- meson being scattered downwards to the left, while the π^+ goes upwards and soon decays via $\pi^+ \rightarrow \mu^+ + \nu$ (ν is a neutrino). The μ^+ meson in turn decays via $\mu^+ \rightarrow e^+ + \nu + \tilde{\nu}$. The top strongly curved track is that of the positron e^+.

Ionization of the atom is thus accompanied by localization of the microparticle in the wave packet of (21) so that ionization may be considered as transition of the particle from the state $\exp(iPQ_0/\hbar)$ at $t_0 = 0$ to the state $\phi_n(Q_1)$ at $t = t_1$ (the notion is the same as that employed previously to characterize points on the locus of the microparticle). The packet of (21) for $t \gg 1/\Omega_{gg'}$ spreads out until the next ionization at time t_2; for

$$t - t_2 \ll \frac{1}{\Omega_{P'P}}$$

we again get a packet of the form of (21), and so on.

In this example the trajectory is described by a series of excitations of heavy atoms A, which lead to localizations of the microparticle, with subsequent initiation of processes in the detector D.

The example is highly schematic, but its significance is very general: the individual history of a particle is expressed as a sequence of macroscopic events.

Figure 13 shows the history of a π meson which collides with a proton and so generates two new π mesons. The meson tracks are formed by series of bubbles in liquid hydrogen. Could we not hope for a more detailed description of the history of a single macroparticle?

Might we in such a description avoid the chain of macroscopic events, i.e. the language of microcosmical catastrophes in which the microparticle tells us its history? For each bubble in the bubble chamber reflects a catastrophe in the microcosm.

It may seem that there in no hope in the search for such a more detailed description of the history of a microparticle; in fact, we cannot point to a single experimental fact that would indicate that quantum mechanics is incomplete within the range of atomic phenomena of the microcosm.

But, all the same, we should be cautious and remember the quotation from Prutkov: Who interferes with the seeker for unwettable gunpowder?

REFERENCES

1. L. I. Mandel'shtam, *Collected Works* [in Russian], vol. 3, p. 397, Izd. AN SSSR, 1950.
2. E. B. Dynkin, *Markoff Processes* [in Russian], Fizmatgiz, 1963.

SELECTED BIBLIOGRAPHY

1. Whittaker, E.T., *A Treatise on the Analytical Dynamics of Particles and Rigid Bodies*, Cambridge 1937.
2. Sommerfeld, A., *Atombahn und Spektrallinien*, Braunschweig 1951.
3. Born, Max, *Z. Phys.* **153** (1958) 372–388.
4. Madelung, E., *Die mathematischen Hilfsmittel des Physikers*, Springer-Verlag, Berlin, Göttingen, Heidelberg, 1957.
5. Gibbs, W., *Elementary Principles in Statistical Mechanics developed with Especial Reference to the Rational Foundation of Thermodynamics* (Yale Bicentennial Publications), Scribners Sons, New York, 1902.
6. Von Neuman, J., *Mathematische Grundlagen der Quantenmechanik*, Julius Springer Verlag, Berlin, 1932.
7. Bohr, N., *The Library of Living Philosophers: Albert Einstein. Philosopher, Scientist* (1949), p. 201.
8. Dirac, P., *The Principles of Quantum Mechanics*, Clarendon Press, Oxford, 1930.
9. Mott, N.F. and Massey, H.S.W., *The Theory of Atomic Collisions*, Oxford 1949.
10. Heisenberg, W., *The Physical Principles of the Quantum Theory*, Chicago 1930.
11. Pauli, W., in *Handbuch der Physik*, Bd. 24/I, Berlin 1933.
12. Einstein, A., Podolsky, B., and Rosen, N., *Phys. Rev.* **47** (1935) 777.

Made in the USA
Monee, IL
08 July 2026

56665101R00081